UNITEXT for Physics

UNITEXT for Physics series publishes textbooks in physics and astronomy, characterized by a didactic style and comprehensiveness. The books are addressed to upper-undergraduate and graduate students, but also to scientists and researchers as important resources for their education, knowledge, and teaching.

More information about this series at https://link.springer.com/bookseries/13351

Sanjeev Dhurandhar · Sanjit Mitra

General Relativity and Gravitational Waves

Essentials of Theory and Practice

 Springer

Sanjeev Dhurandhar
Inter-University Centre for Astronomy
and Astrophysics (IUCAA)
Pune, Maharashtra, India

Sanjit Mitra
Inter-University Centre for Astronomy
and Astrophysics (IUCAA)
Pune, Maharashtra, India

ISSN 2198-7882　　　　　　　　　ISSN 2198-7890　(electronic)
UNITEXT for Physics
ISBN 978-3-030-92337-2　　　　　ISBN 978-3-030-92335-8　(eBook)
https://doi.org/10.1007/978-3-030-92335-8

This Springer imprint is published by the registered company Springer Nature Switzerland AG
The registered company address is: Gewerbestrasse 11, 6330 Cham, Switzerland

I like the authors' style and the choice of topics. They have taken an unusual approach so that the reader gets to see how relativity works in practical situations. Normally a text in relativity presents the subject as an abstract mathematical exercise. The authors have worked out examples wherein the reader appreciates how the theory works.

Jayant Narlikar

To my parents (especially my mother) and Bhalchandra Dhurandhar (my father's uncle) who indulged my childhood scientific antics.

—Sanjeev Dhurandhar

To my parents and my uncle, who did their best for my education.

—Sanjit Mitra

Preface

Several high-quality textbooks at various levels have been published on General Relativity (GR). Then why, one may ask, write another book? There are two reasons why we chose to write this book. The first reason is that, if there had been a single book available that could have served our purpose, we might not have been motivated enough to write this book! A number of universities and institutions offer a half-semester course on General Relativity to final year Masters' students in Physics or as part of Ph.D. coursework. We have been teaching such courses for years (or even decades), but we could not recommend for such a course a single self-contained book. We had to refer to multiple books for different topics such as special relativity, tensor calculus, differential geometry, black holes, etc. inevitably with different conventions, notations, etc. Often, in order to do justice, the books became too voluminous with interlinked concepts which were difficult to disentangle. It is not uncommon to encounter students who have completed a course on GR without having any insights into the rationale/prescription for parallel transport in Riemannian geometry. Referring students to advanced texts has the consequence of bogging them down and making them lose sight of the main objective of learning general relativity. We have, therefore, adopted a middle-ground, in which the contents can be followed through smoothly without omissions. The instructor may teach a short course without compromising on the essential mathematical concepts which provide a reasonably solid foundation. This is our main motivation in writing this book. We expect that students will find it easy to follow the introductory topics covered and gain valuable insights from this book. Moreover, a clear understanding of the basics will also empower the students to ask advanced questions and further seek answers from advanced texts.

The other major motivation is to provide an introduction to gravitational waves (GW). With the detection of GW and given the ever-increasing sensitivity of GW detectors, the field is sizzling. A large number of researchers have been motivated to enter into this field to carry out research in the observational aspects. Again there is no short introduction readily available that can be understood in a short time without sacrificing content. In fact, fundamentals on GW data analysis can rarely be found in introductory books. Moreover, many of these researchers may not be familiar with GR either, but we believe that they would appreciate the field more if they are provided with a basic introduction not requiring too much effort. This strongly motivated us to put significantly more emphasis on GW and make it an essential one-stop introduction package for someone beginning a research career in GW.

The authors would like to thank Jayant Narlikar and Tarun Souradeep for their encouragement. We are grateful to our students and younger colleagues with whom we have held animated and spirited discussions from which the book has significantly benefited.

Finally, SVD would like to thank NASI, India, for the Platinum Jubilee Fellowship and SM would like to acknowledge DST, India, for the Swarna Jayanti Fellowship. A part of this book was written during both fellowships.

Pune, India   Sanjeev Dhurandhar
                                                                                              Sanjit Mitra

Contents

Chapter 1
Overview of Special Relativity

In this chapter we give a short review of Einstein's special theory of relativity.

1.1 Introduction

Why relativity? (Griffiths [2012]) Historically, classical mechanics which involved macroscopic objects moving with velocities small compared to the speed of light was well explained by the Newton's laws of motion and Galilean relativity. Newton's laws were form-invariant under the Galilean transformations. We consider an inertial frame O (a frame in which Newton's laws hold) described by the Cartesian coordinates (x, y, z). Suppose an inertial frame O' with coordinates (x', y', z') moves with uniform velocity v with respect to the inertial frame O and say we orient the frames so that the motion is along the x-axis with their origins coinciding at $t = 0$, then the coordinates in the two frames are connected by the equations:

$$x' = x - vt, \quad y' = y, \quad z' = z. \tag{1.1.1}$$

Newton's laws retain their form under these transformations. Note that nobody talks of transforming the time coordinate—it is absolute! This means if we dare write down t' as the time in the moving frame O', then $t' = t$, where t is the time in the frame O. So as long as one remained inside one's limits of applicability (such as velocities of material bodies small compared to the speed of light) this framework was sufficient. This is the principle of relativity—laws of physics are the same in all inertial frames. Since at that time only classical mechanics was the physics, these transformations

---

The original version of this chapter was revised: Belated corrections in equations have been updated. The correction to this chapter is available at https://doi.org/10.1007/978-3-030-92335-8_10

© The Author(s), under exclusive license to Springer Nature Switzerland AG 2022, corrected publication 2022
S. Dhurandhar and S. Mitra, *General Relativity and Gravitational Waves*, UNITEXT for Physics, https://doi.org/10.1007/978-3-030-92335-8_1

sufficed. So one singled out a family of frames, all moving with uniform velocities with respect to each other—the inertial frames—in which Newton's laws were true.

The problem arose when *electromagnetism* and with it the corresponding Maxwell's equations came on the scene. Maxwell's equations are differential equations describing the electromagnetic fields denoted by **E** and **B**. This was also another part of physics—new physics at that time. The principle of relativity must also encompass this added physics—Maxwell's equations must also obey the principle of relativity and therefore transform form covariantly under Galilean transformations. This turned out not to be so. So it seemed either Maxwell's equations were incorrect or the principle of relativity had to be abandoned. But it was time and again found that nothing was wrong with Maxwell's equations and also the principle of relativity which was so philosophically deep in nature and all encompassing was hard to abandon. Einstein solved the problem by not abandoning either of these but by completely revamping the ideas of space and time—space is that which is measured by rulers and time is that which is measured by clocks. One must therefore give up on preconceived ideas about space and time such as the absolute nature of time and simultaneity. Thus time and simultaneity fell from their high pedestal and became quantities relative to a given inertial frame or an inertial observer. So it was the Galilean transformations connecting two inertial frames in relative motion that were incorrect and would have to be replaced by a different set of transformations—the Lorentz transformations. The reason Galilean transformations "work" in everyday life is because one is usually dealing with objects which travel much slower than the speed of light. In this case the Lorentz transformations approximate the Galilean transformations to a high degree of accuracy. However, in this age of particle accelerators and the GPS etc., either we are dealing with particles travelling with speeds close to that of light or making very precise time measurements of objects in relative motion in different gravitational fields. Then Galilean transformations give inaccurate or inadequate results and we are forced to give these up and apply new principles and concepts; those of relativity.

We will now first state the two principles on which the special theory of relativity (special because we are dealing with physics in the absence of gravity—the theory where gravity is also part of the physics is called the general theory of relativity) is founded. Then we will go on to derive some standard results of the theory. We will however, do this geometrically as much as possible, first because, a picture explains the physical situation at a glance much better and secondly because this book is on the general theory of relativity—the theory is inherently geometrical in nature.

1.2   Postulates of Special Relativity (SR)

The following are the two postulates of SR:

1. The laws of physics are the same in every *inertial* reference frame.
2. Light has the same speed in vacuum denoted by $c = 2.99792458 \times 10^8$ metres/second in every inertial frame.

The second postulate requires some elaboration. Light as such has nothing to do with relativity—light just happens to travel with this universal speed c and in general is the speed of zero rest mass particles like photons. Any other zero rest mass particle such as the graviton would also travel with the same speed. In fact relativity goes beyond electromagnetism; it does not refer to any specific physical interaction such as electromagnetism, but concerns the nature of space and time in which the physics takes place. c also happens to be the maximum speed with which particles can travel. It is also the second postulate which leads to apparent "paradoxes" in relativity—*there are no real paradoxes in relativity because it is a consistent theory.* The so-called paradoxes occur because of wrongly mixing up Newtonian and relativistic concepts.

Consider two nearby events whose coordinates in frame O are given by (ct, x, y, z) and $(c(t + dt), x + dx, y + dy, z + dz)$. Since c is independent of reference frames, one can see that the quantity

$$\mathrm{d}s^2 := c^2\mathrm{d}t^2 - \mathrm{d}x^2 - \mathrm{d}y^2 - \mathrm{d}z^2, \tag{1.2.1}$$

is zero for light travelling in vacuum in any frame. In fact one may use symmetry arguments and the principle of relativity to show that even in the case when $\mathrm{d}s^2 \neq 0$ for arbitrary nearby events, the quantity ds^2 must be invariant in special relativity (See e.g., Landau and Lifshitz [1980]). That is,

$$\mathrm{d}s'^2 := c^2\mathrm{d}t'^2 - \mathrm{d}x'^2 - \mathrm{d}y'^2 - \mathrm{d}z'^2 = c^2\mathrm{d}t^2 - \mathrm{d}x^2 - \mathrm{d}y^2 - \mathrm{d}z^2 =: \mathrm{d}s^2 \tag{1.2.2}$$

One can go even further and generalise to non-infinitismal intervals for well separated events. Now the coordinate separations are denoted by $c\Delta t, \Delta x, \Delta y, \Delta z$ instead of cdt, dx, dy, dz. We denote the finite interval by Δs which is also invariant when transforming from one inertial frame to another. The transformations which keep this form of the interval invariant are precisely the Lorentz transformations which relate (ct, x, y, z) to (ct', x', y', z'). Later we will present the explicit form of the Lorentz transformations.

An important quantity is the proper time elapsed during the trajectory of a particle from say a point A to point B which lie on its trajectory. One can now notice the fundamental difference in SR from Galilean relativity, that time is no longer an invariant quantity (that is, dt and dt' need not be the same), each frame has its own clock. In the rest frame of a particle, only time changes in its own clock, and so the interval for infinitesimally separated events on its trajectory, called the world-line, the time interval is $\mathrm{d}\tau := |\mathrm{d}s^2|^{1/2}/c$, appears to be *maximum*. This is called the "proper time interval", as the time measured by a clock is the "proper time" in its own (rest) frame.

The proper time is related to the time interval measured in any other frame through the following relation

$$\mathrm{d}\tau = \left[\mathrm{d}t^2 - \frac{(\mathrm{d}x^2 + \mathrm{d}y^2 + \mathrm{d}z^2)}{c^2}\right]^{1/2} = \sqrt{1 - v^2(t)/c^2}\,\mathrm{d}t, \tag{1.2.3}$$

where $v(t)$ is the speed of the particle. Clearly, the interval between the same two events, say the period of a pendulum or the interval between ticks of the same clock, the reading on the clock attached to a frame $\mathcal{O}$, $d\tau$, is always smaller than the interval dt measured from a frame $\mathcal{O}'$ which is moving with respect to the events. That is **moving clocks run slow**.

In relativity, the factor $1/\sqrt{1 - v^2(t)/c^2}$ appears time and again, and so we use a dedicated symbol for this factor,

$$\gamma(t) := \frac{1}{\sqrt{1 - v^2(t)/c^2}}. \tag{1.2.4}$$

In this notation $d\tau = dt/\gamma(t)$. In the **non-relativistic limit**, $|v| \ll c$, $\gamma = 1$, so the proper time must be close to the coordinate time interval, $d\tau \approx dt$. $\gamma(t)$ is called the Lorentz factor. Eq.(1.2.3) can be integrated from A to B to arrive at the proper time τ_{AB} elapsed from A to B:

$$\tau_{AB} = \int_{t_A}^{t_B} \frac{dt}{\gamma(t)}. \tag{1.2.5}$$

From the foregoing a question may arise. If $\mathcal{O}$ and $\mathcal{O}'$ are in relative motion, then $\mathcal{O}'$ finds $\mathcal{O}$'s clocks running slow, but by the same token, by the principle of relativity, $\mathcal{O}$ will find $\mathcal{O}'$'s clocks running slow! How can both be right? Is the theory of relativity inconsistent? That this does not contradict the principle of relativity can be shown using a space-time diagram (see Exercise 1.2).

1.3 Space-Time Diagrams: A Picture Is Worth a Thousand Words

For simplicity, we ignore the y and z directions and consider only two dimensions, one spanned by the x-axis drawn horizontally and the other by the time axis ct drawn vertically. Note that we have multiplied the time coordinate t by the factor c—the speed of light—so that both the axes are in the same units, namely, length. It is easy to add in the other space axes later and deduce consequences. But for most purposes two dimensions suffice when the physical situation is simple and we have the freedom to align our axes.

Now consider an event which for definiteness sake occurs at the origin (we can always shift the origin to the event site otherwise). Now let us explore the relation of different regions of the space-time with respect to this event with the help of the diagram shown in Fig. 1.1.

The space-time points naturally get divided into disjoint regions according as the quantity $\Delta s^2 = c^2 \Delta t^2 - \Delta x^2$ is positive, zero or negative. If $\Delta s^2 > 0$, the separation between the event (ct, x) and the event O at the origin—we take the reference point to be at the origin— is called time-like, when the 'time part', namely $c^2 \Delta t^2$ dominates over the 'space part' Δx^2. Physically, for a time-like interval, utilising the invariance of interval, one can find a (primed) frame where these events are at the same point

Fig. 1.1 A space-time
diagram showing past, future
and causally unconnected
space-like events

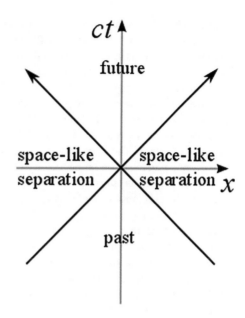

in space, that is, $\Delta x' = 0$ and $\Delta s^2 = c^2 \Delta t'^2$. By a similar reasoning such events could never be simultaneous (why not?). All the points satisfying $\Delta s^2 > 0$ occupy the 'inside' of a double cone as shown in the diagram. Then $\Delta s^2 = 0$ defines the lines which are given by the equations $x = \pm ct$. These lines form a boundary to this region and are the lines at slopes of $\pm 45°$ through the origin. Note that this relation between pairs of events is *observer independent* simply because Δs^2 is an invariant, that is, its numerical value remains the same for all observers. When $\Delta s^2 < 0$, the separation between events (ct, x) and O is termed spacelike. One can find a frame in which the separation is purely in space, that is, $x' \neq 0$ but $ct' = 0$. The latter also implies that the events are simultaneous in this frame. These are the regions outside the boundary defined by the lines $x = \pm ct$. So this boundary has a special significance of demarkating the various regions and are known as *light lines* (in two dimensions). This is because these are the worldlines of light or photons.

Two immediate generalisations are in order: (i) We can relax the condition on dimensions and include also the y and z coordinates; (ii) secondly, one may shift O to a general position (ct_1, x_1), so that now the spacetime intervals are between the two events (ct_1, x_1) and (ct_2, x_2), namely, $\Delta s_{12}^2 = c^2(t_2 - t_1)^2 - (x_2 - x_1)^2$. Considering first only the first generalisation (i), $\Delta s^2 = 0$ gives the equation:

$$c^2 t^2 = x^2 + y^2 + z^2. \tag{1.3.1}$$

If one suppresses the z coordinate, effectively setting $z = 0$, and draws only the ct, x, y axes, then the above equation becomes $c^2 t^2 = x^2 + y^2$, which geometrically represents a right circular cone. Including the z coordinate one gets what is called a hypercone (a 3-dimensional space embedded in 4 dimensions), which is difficult (in fact impossible) to visualise. We call this the light-cone and is rigorously defined through the equation Eq. (1.3.1). We will also use the word lightcone instead of light lines even for the two dimensional case without any cause for confusion.

The inside of the light cone is naturally divided into two disjoint regions, according as the time coordinate ct is positive or negative, so that we have an upper cone and a lower cone. This means that the events in the upper cone are in the future and the lower cone are in the past, as physical trajectories always follow the arrow of time from past to future. The upper cone is called the future light cone and consists of those events which can be influenced by O, while the lower cone is called the past lightcone and consists only of those events which can influence O. Thus the lightcone implies a causal relationship of events in the two cones with respect to O.

We can now go to the second generalisation where now instead of the origin, the reference point is an arbitrary fixed point (ct_1, x_1, y_1, z_1). Both generalisations are possible at the same time and one can construct a light cone at any arbitrary point (ct_1, x_1, y_1, z_1) in the spacetime. More generally, one can construct light cones at every point in the spacetime. This is called a light cone "field" on the spacetime—a lightcone is attached to each point of the spacetime. Since the lightcones imply causal relationship between events, the lightcone field determines the causal structure of the spacetime. The concept of a blackhole emerges naturally from the causal structure of a spacetime as we will see later in Chap. 6.

1.3.1 Setting Up the Axes for a Moving Observer

1.3.1.1 Slopes of the Axes

Let an observer O' move with respect to O with uniform velocity v along the x-axis. We already have set up O's axes namely, (ct, x). Our job now is to set up O' axes which we denote by (ct', x'). To make the bookkeeping simple, let their origins coincide—that is, the same point which we denote by O has coordinates $(0, 0)$ in both axes. We have now two steps to take (i) determine the slope of ct' and x' axes, (ii) calibrate each of the axes—if we are using metres as a unit of length, then where are the points 1 metre, 2 metres, etc. on the x' axis and similarly, we must mark analogously on the ct' axis.

Getting the slope of the ct' axis is simple, because it is the worldline of the observer O' and is described by $x = vt$. Or writing the same in the (ct, x) diagram, the same equation can be written in the form:

$$ct = \frac{c}{v}x . \tag{1.3.2}$$

Thus the ct'-axis is a straight line with slope c/v drawn through the origin. So now we have one axis. We need to get the x' axis. This requires more work (Schutz, 1995).

Fig. 1.2 Setting up the x'-axis. Note that since the equation of x'-axis is $ct' = 0$, it represents all points that are simultaneous with the origin in O''s frame

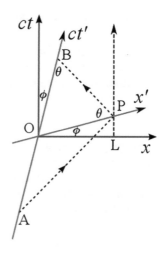

Consider the Fig. 1.2. Let O$'$ emit a radar signal at $ct' = -a$ labelled A, which is so reflected at P that it returns to O$'$ at $ct' = +a$ labelled B. Now it is reasonable for O$'$ to assign a coordinate $ct' = 0$ to the event P. Taking a as a parameter and varying it, the locus of all such events P with $ct' = 0$ would be exactly defined as the x'-axis. Using the fact that AP and PB are at $\pm 45°$, the x'-axis can be geometrically reconstructed in Fig. 1.2 as follows:

Since as shown above, ct' axis has slope c/v, the angle between ct and ct' axis is $\phi = \tan^{-1} v/c$. Since PB makes an angle $45°$ with the ct-axis, the angle OBP which we call θ is $45° + \phi$. Since AO = OB and APB is a right angle, one can draw a circle with AB as the diameter passing through P, so that we have OB = OP = OA and triangles OBP and OAP are isosceles. Hence the angle BOP is $180° - 2\theta = 90° - 2\phi$. Thus from the geometry of the figure one can easily convince oneself that the angle between x' and x axes is ϕ which is the same as that between ct and ct' axes. Thus x'-axis has slope v/c and is given by the equation:

$$ct = (v/c)x . \tag{1.3.3}$$

The x'-axis consists of all points simultaneous with the origin in the frame of O$'$. Eqs. (1.3.2) and (1.3.3) can be directly derived from the Lorentz transformations by setting t' and x' equal to zero.

Note that since this axis is inclined to the x-axis, t is not constant along x'-axis so these points are not simultaneous in the frame of O. Thus which events are simultaneous depends on the observer and is not an absolute concept as is the case in Newtonian mechanics. This concludes step (i).

1.3.1.2 Calibration of the Axes and Invariant Hyperbolae

The loci of constant intervals where one (reference) event is at the origin, trace out
hyperbolae in the space time-diagram (this is in $1 + 1$ dimensions; the situation is
little more complicated in the higher dimensions), given by

$$c^2 t^2 - x^2 = \text{const.} \tag{1.3.4}$$

The hyperbolae are very useful for providing the caliberation of "primed" frame.
They mark equal "distances" or spacetime intervals on the axes, just as circles $x^2 +
y^2 = $ const. do in the usual Euclidean two dimensional space or the plane.

Let us now caliberate the x'-axis. Suppose one wishes to mark the point 1 metre
on the x'-axis, all one needs to do is to draw a hyperbola $x^2 - c^2 t^2 = 1$. Actually
there are two hyperbolae, one chooses the one on the side where x is positive. Now
one finds where it intersects the positive side of the x'-axis and marks this point, say
P. Since $x'^2 - c^2 t'^2 = x^2 - c^2 t^2$ we have at P, $x'^2 - c^2 t'^2 = x^2 - c^2 t^2 = 1$, and since
additionally being the x'-axis, $ct' = 0$, we must have $x'^2 = 1$ at P or since x' has been
chosen to be positive $x' = 1$. This is the 1 metre mark on the x'-axis. Considering
the hyperbola on the negative half plane we could have a point P' which should be
marked $x' = -1$ (not shown in the figure). In this way we can mark the entire x' axis
at a given length l by choosing the appropriate hyperbola $x^2 - c^2 t^2 = l^2$.

A similar procedure can be carried out to caliberate the ct' axis. To mark $+l$
metres on this axis consider the intersection of the relevant hyperbola (in the $ct' > 0$
half-plane) $c^2 t^2 - x^2 = l^2$ with the ct' axis. The intersection gives the required point
say Q on the axis. The procedure is evident from the Fig. 1.3.

Fig. 1.3 The figure shows
invariant hyperbolae which
can be used to caliberate the
primed axes. The
$x^2 - c^2 t^2 = 1$ hyperbola
intersects x' axis at P. This
marks 1 metre on the x' axis.
In this way one may continue
to caliberate the full x' axis
and in a analogous way the
ct' axis

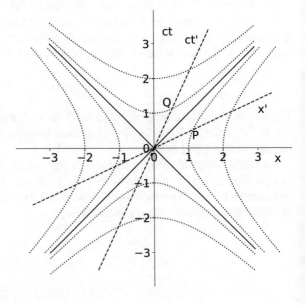

An immediate use of such an exercise is to understand length contraction and time dilation effects for moving observers. These phenomena become evident from the geometry of the spacetime diagram and the calibration of the moving frame. We do this in the next subsection.

1.3.2 Lorentz Contraction and Time Dilation

The fact that simultaneity is not an absolute concept has the consequence of the length measurement, say of a rod, depends on the motion of the observer. A moving observer measures reduced length of the rod. This is known as Lorentz contraction. This effect is opposite to that of time measurements. Time appears dilated in a moving frame. This is the direct consequence of the fact that dt^2 and dx^2 appear with opposite signs in the metric. We treat both of these phenomena geometrically here.

Let us consider a rod of length l stationary in the frame of $\mathcal{O}$ lying along the x-axis with one end coinciding with the origin and the other at L at $t = 0$. In Fig. 1.4, the rod lies along OL at $t = 0$. The rod (considered one dimensional) therefore traces out a world-sheet which is made up of world-lines of each point on the rod. The world sheet is described by the region $0 \le x \le l$ and ct taking up all values. This is the history of the rod for all times and is shown by the shaded region in the figure. Now consider the observer $\mathcal{O}'$. In order to measure the length of the rod, he must look at both ends of the rod at the same time *in his frame*. This is where simultaneity comes in. The events for $\mathcal{O}'$ simultaneous with each other lie along lines parallel to the x'-axis. In particular, if we consider the x'-axis itself, these events are simlutaneous for $\mathcal{O}'$—infact they occur at $ct' = 0$. Thus $\mathcal{O}'$ measures OL' as the length of the rod, which is a different cross-section of the worldsheet and hence will in general measure a different length say OL' $= l'$. How do l and l' relate? To see this draw a hyperbola through L and let it cut the x'-axis at M say. Now by the preceding arguments of invariant intervals, OM $= l$, because L and M lie on the same hyperbola. Clearly, OM > OL'. Thus $l > l'$ and we have a contraction—Lorentz contraction. The next question is how much is this contraction? For this we do some geometry.

One way is to use coordinate geometry. We need to obtain the spacetime interval OL'. For this we must obtain the coordinates of L' in the (ct, x) frame. We notice that L' is the intersection of the line $x = l$ with the x'-axis. So the x coordinate of O' is just l. The ct coordinate is obtained from the equation of the x'-axis: $ct = (v/c)x$. But this immediately yields the ct coordinate of L', namely, $(v/c)l$. Thus the coordinates of L' are $((v/c)l, l)$. Thus the proper length l' of the interval OL' is given by $l'^2 = l^2 - (v/c)^2 l^2 = l^2(1 - (v/c)^2)$. From this we immediately have:

$$l' = l\sqrt{1 - \frac{v^2}{c^2}}. \tag{1.3.5}$$

The geometry of such a space with an indefinite metric is known as Lobachevskian or simply hyperbolic geometry. Now in this geometry we have a theorem analogous

Fig. 1.4 Lorentz contraction
and time dilation: The
vertical lines through O and
L parallel to the *ct* axis are
the ends of the rod whose
length is *l* in $\mathcal{O}$'s frame
which is also its rest frame.
$\mathcal{O}'$ considers the
cross-section OL′ to be the
rod which he measures as
$l' < l$. Similarly, for the time
elapsed $T =$ OQ in O's
clock, $\mathcal{O}'$ measures the
corresponding time $T' =$
OQ′ because Q′ is
simultaneous with Q. From
the invariant hyperbola, OQ
= OS =T. From the figure
OQ′ > OS and so $T' > T$.
Time is dilated

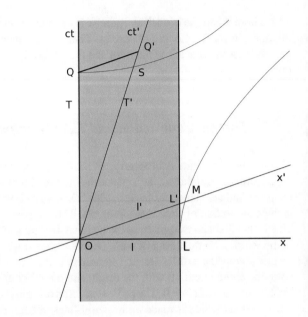

to Pythagoras: the square of the hypotenuse is now the *difference* of the squares of the
two sides. We can apply this modified Pythagoras theorem. Consider triangle OLL′.
The angle OLL′ is a right angle. whence we get $OL'^2 = OL^2 - LL'^2$. Now LL′ is
just $(v/c)L$, since the slope of the x'-axis is v/c. Putting all this together gives us
$l'^2 = l^2(1 - (v/c)^2)$ as derived earlier leading to the relationship described by Eq.
(1.3.5).

Time dilation is just the flip side of the coin . Let $\mathcal{O}$ carry a clock whose world line
is the ct-axis. Consider a time interval from O to Q, which we call T. This is the time
measured in $\mathcal{O}$'s frame. What does $\mathcal{O}'$ measure? To see this, draw a parallel line to the
x'-axis through Q. Then all the events on this line are simultaneous as seen by $\mathcal{O}'$. In
particular, the event Q′, which is the intersection of the ct'-axis with the parallel line
through Q, is simultaneous with Q. Thus $\mathcal{O}'$ measures the time interval equal to OQ′.
Let us call this time interval as T′. Now how does T′ compare with T? To see this draw
an invariant hyperbola through Q. It cuts the ct axis at S say. By the definition of the
invariant hyperbola, the spacetime interval OS is T. From the diagram, S lies below
Q′, and so $T' > T$. Thus time is dilated. By how much? This question can be answered
as before by either using coordinate geometry or just geometric arguments employing
hyperbolic geometry. We use the latter approach and leave the coordinate approach
as an exercise to the reader. Consider the triangle OQQ′. Note since QQ′ is parallel to
the x'-axis it is also orthogonal to the ct' axis—again the indefinite metric (although
it does not 'look' orthogonal because of our Euclidean prejudice). Thus QQ′O is
the right angle here. (We could have drawn the primed axes as right angled and the
unprimed as inclined). Thus we have the equation $OQ^2 = OQ'^2 - QQ'^2$. Analogous
argument to the one above used in Lorentz contraction, shows that $QQ' = (v/c)cT'$

or $c^2 T^2 = c^2 T'^2 - (v/c)^2 c^2 T'^2$ which gives us the result:

$$T' = \frac{T}{\sqrt{1 - \frac{v^2}{c^2}}} .$$

(1.3.6)

This completes our discussion.

1.4 Lorentz Transformations

1.4.1 The Lorentz Transformation Equations

We will now relate the coordinates of events measured in two different reference frames moving with respect to each other with uniform velocity (Padmanabhan, 2010). Consider an observer $\mathcal{O}'$ moving with respect to an observer $\mathcal{O}$ on a trajectory given by $x(t) = vt$. For simplicity, the clocks are synchronised in such a way that the origin of the reference frames of $\mathcal{O}'$ and $\mathcal{O}$ coincide, that is, $(ct', x') = (0, 0)$ at $(ct, x) = (0, 0)$, as shown in Fig. 1.5. Our task will be to relate the pairs of coordinates of every other events, (ct, x) and (ct', x'), measured by $\mathcal{O}$ and $\mathcal{O}'$ respectively.

As discussed above, the observer $\mathcal{O}'$ can set up the coordinates by sending and receiving light rays to and from and event, say P. If a light ray sent from A and reflected by the event P is received at B, then, since AP and BP make $\pm 45°$ angle with the x-axis, one may write:

$$ct_P - x_P = ct_A - x_A ,$$

(1.4.1)

$$ct_P + x_P = ct_B + x_B .$$

(1.4.2)

Substituting $x_{A,B} = vt_{A,B}$ and the time dilation relation $t_{A,B} = \gamma t'_{A,B}$ in the above equation, one gets

Fig. 1.5 Illustration of how an observer moving along a trajectory with uniform velocity, $x(t) = vt$, can find the coordinates of an event P

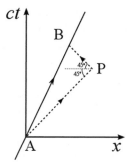

$$ct_P - x_P = \gamma\, ct'_A\, (1 - v/c) \tag{1.4.3}$$

$$ct_P + x_P = \gamma\, ct'_B\, (1 + v/c)\,. \tag{1.4.4}$$

Adding and subtracting the above equations and dividing by two one gets

$$ct_P = \gamma\, [c(t'_B + t'_A)/2 + v(t'_B - t'_A)/2] \tag{1.4.5}$$

$$x_P = \gamma\, [c(t'_B - t'_A)/2 + v(t'_B + t'_A)/2] \tag{1.4.6}$$

Finally, since the coordinates of the event P according to $\mathcal{O}'$ are $t'_P = (t'_B + t'_A)/2$ and $x'_P = c(t'_B - t'_A)/2$ and P is an arbitrary event, one can write in general

$$ct = \gamma\, [ct' + (v/c)\, x']\,, \tag{1.4.7}$$

$$x = \gamma\, [x' + (v/c)\, ct']\,, \tag{1.4.8}$$

which are the famous Lorentz transformation equations. They look more symmetric in the $c = 1$ units:

$$t = \gamma\, (t' + v\, x')\,, \tag{1.4.9}$$

$$x = \gamma\, (x' + v\, t')\,. \tag{1.4.10}$$

These can also be written as a "rotation" in four-dimensions in terms of "rapidity" $\chi := \tanh^{-1} v/c$ [i.e., $\sinh \chi = \gamma(v/c)$ and $\cosh \chi = \gamma$] as

$$x = x'\, \cosh \chi + ct'\, \sinh \chi\,, \tag{1.4.11}$$

$$ct = x'\, \sinh \chi + ct'\, \cosh \chi\,. \tag{1.4.12}$$

One interesting way to look at Lorentz transformation is that (Padmanabhan [2010])

$$(ct' \pm x') = e^{\mp \chi}\, (ct \pm x)\,, \tag{1.4.13}$$

which shows that Lorentz transformation stretches $(ct - x)$ and compresses $(ct + x)$ to keep $(c^2 t^2 - x^2)$ invariant (constant hyperbolae). Incidentally, $u = ct - x$, $v = ct + x$ are known as the null coordinates, $ds^2 = du\, dv$. They play an important role in describing null hypersurfaces and in particular blackholes e.g. the Eddington-Finkelstein coordinates (Misner et al. [1973]).

Above we assumed the relative velocity between two frames to be along the x-axis. This can be done without any loss of generality because we can always orient our axes so that the relative velocity is along the x-axis. We can also argue that in the directions perpendicular to the x-axis, the coordinates do not change. We may assume rotational symmetry about the x-axis, which implies the relations $y' = y$ and $z' = z$. If these relations were not true, say if $y' > y$ then symmetry would imply $z' > z$ violating the invariance of the interval under the Lorentz transformation. Because then $dy' > dy$ and $dz' > dz$ and we would end up with $ds'^2 \neq ds^2$.

For completeness we present the general expression for Lorentz transformation, where $\mathcal{O}'$ is moving with respect to $\mathcal{O}$ with a uniform velocity $\mathbf{v}$. Writing $\mathbf{r}$ as the three dimensional position vector (Goldstein [1980])

$$ct = \gamma\,(ct' + \mathbf{v}\cdot\mathbf{r}'),\tag{1.4.14}$$

$$\mathbf{r} = \mathbf{r}' + (\gamma - 1)\frac{\mathbf{v}(\mathbf{v}\cdot\mathbf{r}')}{v^2} + \gamma\,\mathbf{v}\,t'.\tag{1.4.15}$$

1.4.2 Velocity Addition

Let a particle move with velocity $\mathbf{V}$ in the O frame with coordinates (ct, x, y, z). Then we have $\mathbf{V} = (dx/dt, dy/dt, dz/dt)$. Now consider the frame O' moving along the x-axis with velocity v with coordinates (ct', x', y', z'). What is the velocity of the particle in the O' frame?

We start with the differentials connecting the two frames. Using the Lorentz transformation equations, we write

$$dt' = \gamma\,(dt - \frac{v}{c^2}\,dx),$$
$$dx' = \gamma\,(dx - v\,dt),$$
$$dy' = dy,$$
$$dz' = dz.$$

Dividing the numerator and denominator by dt one gets

$$V_{x'} := \frac{dx'}{dt'} = \frac{V_x - v}{1 - \frac{vV_x}{c^2}}\tag{1.4.16}$$

$$V_{y'} := \frac{dy'}{dt'} = \frac{V_y}{\gamma\left(1 - \frac{vV_x}{c^2}\right)}\tag{1.4.17}$$

$$V_{z'} := \frac{dz'}{dt'} = \frac{V_z}{\gamma\left(1 - \frac{vV_x}{c^2}\right)}\tag{1.4.18}$$

If $\mathbf{V}$ is purely along the x-axis, then $V_y = V_z = 0$ and we get the formula for addition of velocities in this special case:

$$V' = \frac{V - v}{1 - \frac{vV}{c^2}},\tag{1.4.19}$$

where now both v and V are along the x-axis.

1.5　Four-Vector Notation and Covariant Formalism

In relativity, one generally uses a $1 + 3$ coordinate system $x^\mu \equiv (x^0, x^1, x^2, x^3)$. The index μ takes values $\mu = 0, 1, 2, 3$. Here we have set $ct = x^0, x = x^1, y = x^2, z = x^3$. Note that the spacetime indices are written as superscripts. They are not exponents. This notation facilitates writing of expressions, especially those involving sums. The notation is further made compact by the use of the Einstein summation convention. Using this notation and convention, we can then succinctly express the Lorentz "boost" along the x-axis with velocity v as,

$$x'^\mu = \Lambda^\mu_{\;\nu} x^\nu, \tag{1.5.1}$$

where,

$$\Lambda^\mu_{\;\nu} := \begin{pmatrix} \gamma & -v\gamma/c & 0 & 0 \\ -v\gamma/c & \gamma & 0 & 0 \\ 0 & 0 & 1 & 0 \\ 0 & 0 & 0 & 1 \end{pmatrix}. \tag{1.5.2}$$

In Eq. (1.5.1), we have used Einstein summation convention. There is an implied sum over the index ν which takes values 0, 1, 2, 3. So there are 4 terms on the RHS which have been summed together. ν is called the dummy index (it can be replaced by any other index say $\sigma = 0, 1, 2, 3$). The index μ is called the free index and also takes four values. The index μ labels 4 separate equations - index μ taking values 0, 1, 2, 3. And each equation has the sum of four terms on its RHS, the sum over the index ν. Thus we see how convenient this notation is where the equations can be written so compactly.

Any vector V^μ which transforms according to the law:

$$V'^\mu = \Lambda^\mu_{\;\nu} V^\nu, \tag{1.5.3}$$

is called a "four-vector" or more precisely a Lorentz 4-vector. The scalar product between two four-vectors V^μ and U^μ is defined as,

$$(V, U) \equiv \eta_{\mu\nu} V^\mu U^\nu = V^0 U^0 - V^1 U^1 - V^2 U^2 - V^3 U^3, \tag{1.5.4}$$

where

$$\eta_{\mu\nu} := \begin{pmatrix} 1 & 0 & 0 & 0 \\ 0 & -1 & 0 & 0 \\ 0 & 0 & -1 & 0 \\ 0 & 0 & 0 & -1 \end{pmatrix} \equiv \mathrm{diag}(1, -1, -1, -1). \tag{1.5.5}$$

The quantity $\eta_{\mu\nu}$ is called the metric tensor. Note that the time components appear with a positive sign, while the three space components appear with negative sign. This is called the signature of the metric and is written as $(+, -, -, -)$. It is different from the Euclidean metric, where all the signs are positive - such a metric is called positive definite. $\eta_{\mu\nu}$ is called indefinite metric or more precisely, the Lorentz metric. The metric is the key quantity in general relativity. In SR it has the above special form in Cartesian coordinates. Note that all components are either 0 or ± 1 and thus constants. This has special significance in general relativity—the absence of the gravitational field—we will come to this in the latter part of the book. Also using Einstein summation convention, the equation is written in a compact form—there are sixteen terms on the RHS! But here only 4 of them are non-zero, because $\eta_{\mu\nu}$ has just 4 non-zero components.

Covariant formalism is the mathematical way of ensuring that the physical laws retain the same form in every inertial frame. In Newtonian mechanics, when an equation is written in terms of three-vectors to describe a physical phenomena, the mathematical form implicitly ensures the invariance of the form or *covariance* of the equation under the relevant coordinate transformations, namely, the Galilean transformations, which consist of rotations (orthogonal transformations) and translations. For example, the form of Newton's law of universal gravitation, the expression for force exerted by a mass m_1 at $\mathbf{r}_1$ on m_2 at $\mathbf{r}_2$,

$$\mathbf{F}_{21} = G m_1 m_2 \frac{\mathbf{r}_2 - \mathbf{r}_1}{|\mathbf{r}_2 - \mathbf{r}_1|^3}, \qquad (1.5.6)$$

where G is Newton's constant of gravitation, is covariant under the Galilean coordinate transformations, because all the three-vectors, $\mathbf{F}$, $\mathbf{r}_1$ and $\mathbf{r}_2$ are transformed by the same rule.

However, when we switch over to SR, the physical laws should be written in terms of quantities which transform covariantly under Lorentz transformations and not Galilean transformations. These are the Lorentz scalars, vectors and tensors on the background of four dimensional space-time. Physical laws written as equations in these quantities are automatically covariant.

An example will illustrate this point. Consider a small volume element moving with velocity $\mathbf{v}$ with respect to frame O. Then, the charge density ρ_0 in the rest frame of the volume element, appears to be

$$\rho = \gamma \rho_0, \qquad (1.5.7)$$

in O. This is because, although the total charge is conserved, the volume is Lorentz contracted by the factor γ^{-1} and therefore in the moving frame the charge density is increased by the factor γ. The current density, $\mathbf{J} := \rho \mathbf{v}$, is then,

$$\mathbf{J} = \gamma \rho_0 \mathbf{v}. \qquad (1.5.8)$$

This motivates the form of the a current density four-vector $J^\mu := \rho_0 \, dx^\mu / d\tau$, where τ is the proper time measured by a clock attached to the volume element. The time component is $J^0 = c\gamma\rho_0 = c\rho$ of the full 4-vector $J^\mu = (c\rho, \rho\mathbf{v})$. We use units as in Jackson's Classical Electrodynamics (Jackson [1998]). Now J^μ transforms as a four vector. Why? Because dx^μ transforms as a four vector : $dx'^\mu = \Lambda^\mu_\nu \, dx^\nu$, and further

$d\tau$ and ρ_0 are scalars or invariants under Lorentz transformations, so that multiplying or dividing by these quantities does not change the four-vector character of dx^μ. The advantage is that, now we can write the physical law of charge conservation, namely, the continuity equation,

$$\nabla \cdot \mathbf{J} = -\frac{\partial \rho}{\partial t}, \qquad (1.5.9)$$

more succinctly in four vector form:

$$\frac{\partial J^\mu}{\partial x^\mu} = 0. \qquad (1.5.10)$$

Note that, it is not just about writing a single entity J^μ by combining four indices, but it is about making sure that the new quantity actually transforms as a four-vector. In this example, it would not be proper if one just extended the non-relativistic continuity equation to the four-vector form, one would also have to show physically that when viewed from other frames, the components do transform as four vectors. Eq. (1.5.10) shows that the continuity equation is true in every inertial frame.

1.6   Relativistic Mechanics

The above mentioned principles can be also applied to mechanics. In Newtonian mechanics, the physical quantities are scalars, vectors etc. with respect to the the Galilean group of transformations. While in SR one must generalise the corresponding quantities so that they are scalars, vectors, tensors under the Lorentz group of transformations. In particular, the momentum 3-vector transforming under the Galilean group must be changed to a Lorentz 4-vector. Also it must have the property that the space part of this 4-vector must reduce to the Newtonian momentum 3-vector in the limit of $v \ll c$. This programme is carried out below.

Before defining momentum, one must first define velocity. There are two obvious ways of defining velocity, by dividing the spatial separation $\mathbf{dr}$ with coordinate time interval dt or proper time interval $d\tau$. A very nice difference between these two has been given in Griffiths [2012] for the speed of an aircraft:

1. Ordinary velocity:

$$\mathbf{v} := \frac{\mathbf{dr}}{dt} \qquad (1.6.1)$$

   is the ground velocity of the plane, which tells a ground observer when it is going to land. Here both $\mathbf{dr}$ and dt are measured in the ground frame.
2. Proper velocity:

$$\mathbf{u} := \frac{\mathbf{dr}}{d\tau}, \qquad (1.6.2)$$

will tell when the plane is going to land in the frame of the passenger, which is relevant if a passenger is going to feel hungry after landing or not! Here $d\mathbf{r}$ is measured in the ground frame, but the time $d\tau$ is measured in the plane's frame and so is the proper time. Thus the velocity so defined is clearly a hybrid quantity, but it has the advantage that it is easy to transform and can be straightforwardly generalised to 4-vector form. It is this velocity which enters into the definition of the 4-momentum.

The guiding principle in defining relativistic momentum is that the law of conservation of momentum must hold in every inertial frame—this is the sacrosanct principle which guides us. In SR the inertial frames are related through the Lorentz transformations. One can easily verify that the Newtonian definition of momentum violates the above principle and thus must be modified. We show this below.

Let us consider one dimensional collision of two particles with masses m_1 and m_2, which have initial velocities u_1 and u_2 and final velocities v_1 and v_2 all of them, in the x direction. In the Newtonian case, in say frame O, we have,

$$m_1 u_1 + m_2 u_2 = m_1 v_1 + m_2 v_2. \tag{1.6.3}$$

Now consider an inertial frame O' moving with velocity V with respect to O along the x-axis. Then in O' we have the new velocities $u_1' = u_1 - V$, $v_1' = v_1 - V$ and similarly for the second particle. Here we have used the Galilean law of combining velocities. But in the frame of O' we also have:

$$m_1 u_1' + m_2 u_2' = m_1 v_1' + m_2 v_2'. \tag{1.6.4}$$

because the additional term of $(m_1 + m_2)V$ appears on both sides of the equation and cancels out. So it seems everything is fine. But no! This is not the true nature of spacetime—the inertial frames O and O' are connected by *Lorentz* transformations and not by Galilean transformations and so the velocity addition law derived in Sect. 1.4.2 must be used. Thus,

$$u_1' = \frac{u_1 - V}{1 - u_1 V/c^2}, \tag{1.6.5}$$

and so on for the other velocities. We thus obtain the initial and final velocities in the frame O'. Now we find that Eq. (1.6.4) is no more true! See Exercise 1.7.

What is wrong? The reason is that the momentum so defined is not a Lorentz vector. We can remedy the situation by defining the momentum 4-vector:

$$p^\mu = m_0 \frac{dx^\mu}{d\tau}, \tag{1.6.6}$$

where m_0 is the mass of the particle in its rest frame called the rest-mass. Now if we write the momentum conservation equation as:

$$p_1^\mu + p_2^\mu = q_1^\mu + q_2^\mu, \tag{1.6.7}$$

where the p^μ s represent initial momenta of the particles and q^μ s represent the final momenta of the particles, the conservation of momentum is manifest. It holds in any other inertial frame such as O' because this is a vector (more generally tensor) equation with respect to Lorentz transformations and therefore must hold in all inertial frames. But what is the relation of the 4-vector so defined to the momentum defined in Newtonian mechanics? To see this, we just take the non-relativistic limit $v << c$ for the space part of this vector. Since in this limit $d\tau \longrightarrow dt$, the space part of this vector reduces to the ordinary momentum vector in Newtonian mechanics.

The discussion so far pertained to the space part of the momentum 4-vector. What about the time part of the momentum 4-vector, or the component p^0? As we see from the 4-momentum conservation equation, the zeroth component of four momentum is also conserved. The zeroth component of the 4-momentum of a particle with rest-mass m_0 travelling with velocity $\mathbf{v}$ is:

$$p^0 := \gamma \, m_0 c = \frac{m_0 c}{\sqrt{1 - (v/c)^2}}, \tag{1.6.8}$$

where $v = |\mathbf{v}|$ is the speed of the particle. The quantity $p^0/c = \gamma m_0 = m$ was initially referred to as "relativistic mass" by Einstein. $E = p^0 c = \gamma m_0 c^2 = mc^2$ is also known as the relativistic energy of the particle - the famous equation $E = mc^2$ is manifest. Although all this seems a purely theoretical proposition, this has been vindicated by countless number of experiments in high energy accelerators. Thus the 1 + 3 split of the 4-momentum vector gives the relativistic energy and momentum as:

$$E = m_0 \gamma c^2, \qquad \mathbf{p} = m_0 \gamma \mathbf{v}. \tag{1.6.9}$$

Similar considerations also lead to the definition of relativistic four-force as the rate of change of four momentum with respect to proper time,

$$K^\mu = \frac{dp^\mu}{d\tau}. \tag{1.6.10}$$

For a particle with invariant rest-mass m_0, one can write an equivalent of Newton's second law of motion, $K^\mu = m_0 a^\mu$, where $a^\mu = dv^\mu/d\tau$ is the four-acceleration. It is also useful to define the relativistic three-force, defined as the rate of change of spatial three-vector component of the four-momentum, $\mathbf{p}$, defined in Eq. (1.6.9) as a function of coordinate time,

$$\mathbf{f} = \frac{d\mathbf{p}}{dt}. \tag{1.6.11}$$

The importance of these quantities will become evident from the following discussion.

Since $p^\mu p_\mu = m_0^2 ds^2/d\tau^2 = m_0^2 c^2$ is a constant, taking the time derivative,

$$p_0 \frac{dp^0}{dt} = \mathbf{p} \cdot \frac{d\mathbf{p}}{dt} = \gamma m_0 \mathbf{v} \cdot \mathbf{f} .$$

Then substituting $p_0 = p^0 = \gamma m_0 c = E/c$, one can write,

$$\mathbf{f} \cdot \mathbf{v} = c \frac{d}{dt} (\gamma m_0 c) = \frac{dE}{dt} . \tag{1.6.12}$$

Thus, the four-force could be decomposed as,

$$K^\mu = (\frac{\gamma}{c} \frac{dE}{dt}, \gamma \mathbf{f}) . \tag{1.6.13}$$

To develop a covariant formalism, one needs to use four-force. For example, the Lorentz force law for a particle with charge q can be written as (using the convention elaborated below),

$$K^\mu = q F^{\mu\nu} v_\nu . \tag{1.6.14}$$

The spatial part then reads,

$$\frac{d\mathbf{p}}{d\tau} = \gamma q (\mathbf{E} + \mathbf{v} \times \mathbf{B}) . \tag{1.6.15}$$

Since $dt = \gamma d\tau$, once then gets

$$\mathbf{f} = \frac{d\mathbf{p}}{dt} = q (\mathbf{E} + \mathbf{v} \times \mathbf{B}) . \tag{1.6.16}$$

Since an inertial observer measures the velocity of the charge as a function of coordinate time, this form is often useful in practice.

1.7 Covariant Formulation of Electrodynamics

Since the Maxwell's equations are already Lorentz covariant no change at all is required in the framework of electrodynamics (which was not the case for mechanics). So, here we would like to discuss another relevant point of how the theory of electromagnetism blends coherently and smoothly into relativistic concepts. Electric field in one frame may appear as a magnetic field or a combination of electric and magnetic fields in another frame moving uniformly with respect to the first frame. This is most easily understood if we view electromagnetism in 4 dimensions. In 4 dimensions, the electric and magnetic fields form the components of an anti-symmetric tensor $F^{\mu\nu}$ given by:

$$F^{\mu\nu} := \begin{pmatrix} 0 & -E_x & -E_y & -E_z \\ E_x & 0 & -B_z & B_y \\ E_y & B_z & 0 & -B_x \\ E_z & -B_y & B_x & 0 \end{pmatrix}. \qquad (1.7.1)$$

Such a tensor may be given in some frame O. In another frame O′, the electromagnetic fields are given by Lorentz transforming this tensor. Thus if $F'^{\mu\nu}$ is the electromagnetic tensor in O′, then, we have:

$$F'^{\mu\nu} = \Lambda^{\mu}_{\alpha} \, \Lambda^{\nu}_{\beta} \, F^{\alpha\beta}, \qquad (1.7.2)$$

where Λ^{μ}_{ν} is the Lorentz transformation matrix connecting the two frames O and O′. Two copies of this matrix are needed because we are dealing with a second rank tensor (more about tensors in later chapters!). Also Einstein summation convention is implied. Thus Eq. (1.7.2) shows that the electric and magnetic fields are mixed up in a non-trivial way, so that, a purely electric field in one frame can appear as a magnetic field or more likely a combination of both types of field. Thus relativity makes it evident that both electric and magnetic fields are manifestations of the same underlying interaction, namely, electromagnetism. Just as space and time get "mixed up" in relativity, so do electric and magnetic fields.

Consider the situation when O′ moves with velocity v with respect to O along the x-axis. Then the Lorentz transformation matrix is given by Eq. (1.5.2). Then Eqs. (1.7.1) and (1.7.2) give the following transformations between the electric and magnetic fields in the two frames:

$$E'_x = E_x, \quad E'_y = \gamma \left(E_y - \frac{v}{c} B_z \right), \quad E'_z = \gamma \left(E_z + \frac{v}{c} B_y \right), \qquad (1.7.3)$$

$$B'_x = B_x, \quad B'_y = \gamma \left(B_y + \frac{v}{c} E_z \right), \quad B'_z = \gamma \left(B_z - \frac{v}{c} E_y \right). \qquad (1.7.4)$$

The discussion above was general. In order to fix ideas, let us consider a simple situation. Consider a point charge q moving with uniform velocity v along the x-axis in the frame O. The goal is to compute the electromagnetic fields in O due to the motion of this charge. The usual $3+1$ way of obtaining the fields is via the Lienard-Wiechart potentials. However, using the ideas of SR, we can do so easily by transforming the fields via a Lorentz transformation. The prodecure is far less cumbersome and brings out the power in the covariant approach to electrodynamics.

Consider a frame O′ in which the charge is stationary—so O′ moves with velocity v with respect to O along the x-axis. But since the charge is stationary in O′, its electric field is the Coloumb field given by the Coloumb's law and its magnetic field is zero. Writing these fields in O′ as **E′** and **B′**, we have the equations:

$$\mathbf{E}' = q \frac{\mathbf{r}'}{r'^3}, \qquad \mathbf{B}' \equiv 0, \qquad (1.7.5)$$

where $r'^2 = x'^2 + y'^2 + z'^2$. Now r' is the distance of the field point from the charge which lies at the origin of O'. Further let the origins of the two frames coincide at $t = 0$. So the position of the charge in O at time t is given by $x = vt$, $y = 0$, $z = 0$. So for the space coordinates we have the Lorentz transformations: $x' = \gamma(x - vt)$, $y' = y$, $z' = z$, where $\gamma = (1 - v^2/c^2)^{-\frac{1}{2}}$. Now let us write down the x component of the electric field in frame of O. From Eq. (1.7.4), we find that the x component of the electric field does not change, e.g. $E_x = E'_x$. Also we must express the primed coordinates of O' in terms of O's coordinates using the Lorentz transformations. Thus we have:

$$E_x = E'_x = q\frac{x'}{r'^3}, \tag{1.7.6}$$

$$= q\frac{\gamma(x - vt)}{[\gamma^2(x - vt)^2 + y^2 + z^2]^{3/2}}. \tag{1.7.7}$$

This gives the x component of the electric field. To find E_y and E_z we must again apply Eq. (1.7.4) but now with care. Now O' is stationary and O moves with velocity $-v$ with respect to O'. Thus in Eq. (1.7.4), we must interchange the primed quantities with the unprimed ones and also change v to $-v$. (Note this change does not matter to E_x.) With this change, we find that $E_y = \gamma E'_y$ and $E_z = \gamma E'_z$. Hence we have:

$$E_y = q\frac{\gamma y}{[\gamma^2(x - vt)^2 + y^2 + z^2]^{3/2}}, \tag{1.7.8}$$

$$E_z = q\frac{\gamma z}{[\gamma^2(x - vt)^2 + y^2 + z^2]^{3/2}}, \tag{1.7.9}$$

and

$$B_x = 0, \tag{1.7.10}$$

$$B_y = -\gamma\frac{v}{c}E'_z = -\frac{v}{c}E_z, \tag{1.7.11}$$

$$B_z = \gamma\frac{v}{c}E'_y = \frac{v}{c}E_y. \tag{1.7.12}$$

where E_y and E_z have been given in the previous set of equations. The fields have been plotted at $t = 0$ in Fig. 1.6 below.

The electric field is still radial but not spherically symmetric—it is squashed in the direction of the motion. And now a magnetic field appears—the field lines are circles whose centres lie on the x-axis. In fact as $v \longrightarrow c$, the electric field in the direction of motion tends to zero and both the fields are now orthogonal to the direction of motion of the charge.

The covariant formulation of electrodynamics can also be done in terms of potentials. This is most useful when we linearise general relativity, because the ideas used here, especially that of the choice of gauge is very important. The electric and magnetic fields are written in terms of the electrostatic potential Φ and magnetic vector

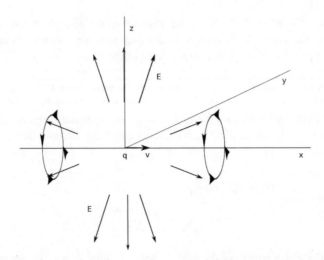

Fig. 1.6 The electromagnetic field of a uniformly moving charge moving along the x-axis. The electric field is squashed in the direction of motion—although radially directed it is not spherically symmetric. The magnetic field circles the x-axis

potential **A** by:

$$\mathbf{E} = -\nabla \Phi - \frac{1}{c}\frac{\partial \mathbf{A}}{\partial t}, \qquad \mathbf{B} = \nabla \times \mathbf{A}. \qquad (1.7.13)$$

Defining the 4-potential vector $A^\mu = (\Phi, \mathbf{A})$, these equations can be written covariantly as:

$$F_{\mu\nu} = A_{\mu,\nu} - A_{\nu,\mu}, \qquad (1.7.14)$$

where the 'comma' denotes partial derivative with respect to the relevant coordinate:

$$\Psi_{,\mu} \equiv \frac{\partial \Psi}{\partial x^\mu}. \qquad (1.7.15)$$

This notation is shorthand and facilitates writing. From the definition of the $F^{\mu\nu}$ tensor in terms of the electric and magnetic fields and the 4-potential A^μ, we can recover Eq. (1.7.13). We must however bring down the indices on the $F^{\mu\nu}$ tensor. This is done via the metric tensor $\eta_{\mu\nu}$ as follows:

$$F_{\alpha\beta} = \eta_{\alpha\mu}\eta_{\beta\nu}F^{\mu\nu}. \qquad (1.7.16)$$

This operation is called 'lowering' the indices. We will do this more systematically later when we do tensors.

The Lorenz gauge condition goes over to:

$$\frac{1}{c}\frac{\partial \Phi}{\partial t} + \nabla \cdot \mathbf{A} = 0 \qquad \longrightarrow \qquad A^{\mu}_{,\mu} = 0. \qquad (1.7.17)$$

After using the Lorenz gauge condition the Maxwell's equations become:

$$\Box A^{\mu} = \frac{4\pi}{c} J^{\mu}, \qquad (1.7.18)$$

where the $\Box$ is the D'Alembertian operator $\partial^2/\partial(ct)^2 - \nabla^2$. Similar technology is used to simplify linearised Einstein's equations, where one chooses a gauge, also called Lorenz gauge by analogy. We will do this when we address weak fields and gravitational waves.

Exercises

1. Use coordinate geometry to derive the expression for time dilation given in Eq. (1.3.6).
   (Hint: See Fig. 1.4. Find coordinates of Q' in O's frame by taking the intersection of the ct' axis with QQ' (note QQ' is parallel to x' axis and passes through $Q(cT, 0)$.)

2. O' moves with uniform velocity v with respect to O along the x-axis. O' says O's clock is running slow. But by the identical argument O finds that O's clocks to be running slow. How can both observers be right? Explain with the help of a spacetime diagram if necessary.

3. Train-tunnel paradox: Let the train and the tunnel have the same proper length say L. The train is moving with uniform velocity v with respect to the tunnel. In the frame of the tunnel, the train is Lorentz contracted and a demon who sits on the top of the tunnel (as demons normally do) drops two shutters at the ends of the tunnel trapping the train. But in the frame of the train the tunnel is Lorentz contracted, so it cannot be trapped! Explain that there is actually no paradox by drawing a space-time diagram of the situation.

4. Transformation of angles: Let O' move with velocity v along the x-axis with respect to O.

   (a) Consider a particle moving with velocity $\mathbf{u} = (u_x, u_y)$ in the $x - y$ plane in the frame O. Let ϕ be the angle its velocity makes with the x-axis, i. e. $\tan \phi = u_y/u_x$. Show that in the primed frame, the corresponding angle ϕ' is given by:

$$\tan \phi' = \frac{\sin \phi}{\gamma(\cos \phi - v/u)}.$$

   (Hint: $\tan \phi' = u'_y/u'_x$; calculate the velocities as in subsection 1.4.2.)

(b) A flash of light is emitted making an angle ϕ with the x-axis in O. Show
that in O' the corresponding angle ϕ' is given by:

$$\cos \phi' = \frac{\cos \phi - v/c}{1 - v/c \cos \phi}.$$

(Hint: write $u_x = c \cos \phi$, $u_y = c \sin \phi$ and use $c \cos \phi' = dx'/dt'$.)

5. Doppler shift: Consider a monochromatic plane electromagnetic wave with frequency ω travelling along the x-axis. Equation of its wave front is $ct = x$ which is
a plane for each time instant t. Let O$'$ move with uniform velocity v with respect
to O along the x-axis. Then the phase of the wave $\phi = \omega t - kx = k(ct - x)$
is Lorentz invariant, that is, $\phi = k(ct - x) = k'(ct' - x')$, where k, k' are wave
numbers in O and O$'$ respectively. Here $\omega = kc$, $k = 2\pi/\lambda$, $k' = 2\pi/\lambda'$. Lines
of constant phase $\phi = -2\pi$, 0, 2π are shown in the Fig. 1.7 below:
The line $\phi = -2\pi$ intersects the x-axis at P. Since the x axis is $ct = 0$, we must
have $OP = \lambda$. Similarly, the same line intersects x'-axis at P' and so $OP' = \lambda'$.
Using the equations of the lines PP' and the x' axis, show that in O's coordinate
system, P' has the coordinates $\lambda/(1 - v/c)(1, v/c)$. Hence compute the proper
length of $OP' \equiv \lambda'$. From this show that:

$$\frac{\lambda'}{\lambda} = \sqrt{\frac{1 + \frac{v}{c}}{1 - \frac{v}{c}}}.$$

6. Superluminal velocities: It is observed in astronomy that some astronomical
objects appear to move faster than the speed of light. How is this possible? We
remark that these are just *apparent* velocities obtained by multiplying the angular
velocity of the object by its distance (the luminosity distance D_L). Then it can be
shown that, if the object is moving at an angle θ with the line of sight and at a
velocity v, then the apparent velocity V is given by,

$$V(\theta) = \frac{v \sin \theta}{1 - (v/c) \cos \theta}.$$

Show that there exists a range of velocities $v < c$ and angles θ for which V can
appear to be greater than c.

7. Momentum conservation in an elastic collision as given in Eq. (1.6.3) with the
Newtonian definition of momentum is not compatible with Lorentz transformations:
Consider two particles of equal mass m travelling along the x-aixs with velocities
$\pm u$ in frame O. Their total momentum $mu + (-mu)$ is zero in O. Now consider
the same system from the point of view of an observer O' moving with velocity v
along the x-axis. In this frame the total momentum is $-2mv$ with the Newtonian
definition. Show explicitly using the velocity addition formula in SR, that the

momenta of the two particles do not add up to $-2mv$, thus violating momentum conservation.

8. Show explicitly that in a collision if energy and momentum are conserved in one inertial frame (lab frame), an elastic collision, then they are conserved in any other inertial frame.

(a) Consider a collision in one space dimension x. Take another frame O' to be moving along the x-axis with velocity $V = \tanh \chi$ with respect to frame O. More specifically, let u_1, u_2 be the velocities with rapidity parameters α_1, α_2 of particles 1 and 2 before collision and v_1, v_2 velocities with rapidity parameters β_1, β_2 after collision respectively in frame O. Let all primed quantities denote the corresponding ones in the O' frame. Then energy momentum conservation in O is given by:

$$m_1 \cosh \alpha_1 + m_2 \cosh \alpha_2 = m_1 \cosh \beta_1 + m_2 \cosh \beta_2$$

$$m_1 \sinh \alpha_1 + m_2 \sinh \alpha_2 = m_1 \sinh \beta_1 + m_2 \sinh \beta_2$$

Show that the above equalities hold also for primed quantities $\alpha_1' = \alpha_1 + \chi$, etc showing that energy and momentum are also conserved in frame O'. (Hint: Use addition formulae for hyperbolic functions).

(b) Use the expressions in (a) to show that the relativistic momentum is conserved in the previous Exercise 1.7.

Fig. 1.7 The figure shows lines of constant phase $\phi = \omega t - kx$ for an electromagnetic wave travelling in the x-direction

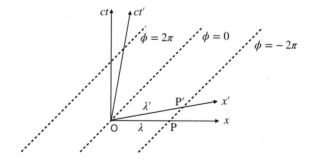

9. For a particle of mass m_0, moving under the influence of a constant three-force $\mathbf{f} = d\mathbf{p}/dt$ as measured from an inertial frame (e.g., a charge moving in a constant electrostatic field), show that the velocity $\mathbf{v}$ as a function of time t starting from rest at $t = 0$ can be expressed as: $\mathbf{v}(t) = \mathbf{f}t/\sqrt{m_0^2 + f^2 t^2/c^2}$. Verify that this is consistent with Newton's laws of motion at low velocity and limits the maximum velocity of the particle to c.

10. Obtain the fields of a uniformly moving charge q by Lorentz transforming the vector potential A_μ. Orient the axes so that the charge moves along the x-axis with velocity v and go to the frame O' in which the charge is stationary and

located at the origin. In O' we have $\Phi' = q/r'$ and $\mathbf{A}' = 0$, where r' is the radial coordinate in O'. Lorentz transform A'_μ back to the O frame.

(a) Show that:

$$\phi(\mathbf{r}, t) = \frac{q}{[(x - vt)^2 + (y^2 + z^2)(1 - v^2/c^2)]^{\frac{1}{2}}},$$

$$A_x(\mathbf{r}, t) = \frac{v}{c}\phi(\mathbf{r}, t), \quad A_y = 0, \quad A_z = 0.$$

(b) Differentiate the potentials obtained above to arrive at the electromagnetic fields $\mathbf{E}$ and $\mathbf{B}$. Use Eq. (1.7.13).

Chapter 2
The Equivalence Principle

The special theory of relativity successfully provided a framework for formulating physical theories in inertial frames which were in uniform relative motion with respect to each other. Physical laws were form-invariant or covariant when transforming from one inertial frame to another. Classical mechanics and electrodynamics were included in this framework at the inception of SR. However, Newton's law of gravitation stood aloof from this—it did not fit in this framework, that is, it was not consistent with SR. Gravity as described by the inverse square law of Newton is instantaneous. The force between two point masses depends only through the distance between the instantaneous positions of the masses at the *same time*. This statement made sense in Newton's theory because there was an absolute time and so the notion of two events being simultaneous was absolute. But in SR, this statement is ambiguous because simultaneity is relative to an observer and so what one observer regards as instantaneous is not necessarily so for another observer moving with respect to the first. Also a displacement of one body can change the force on the other instantaneously, disobeying the limit on the speed of maximum information propagation which is finite and equal to c. Einstein spent a decade to address these issues and came up with one of the most beautiful theories of physics—*the general theory of relativity* (GR). The key idea was to generalise to non-inertial frames—not just frames moving with uniform velocity with respect to each other—frames could now be accelerated with respect to each other. Instead of Lorentz transformations any general transformation could be made and the requirement was that laws of physics should be covariant under these general transformations. The general nature of the transformations connecting the frames led to the nomenclature—general relativity. But general transformations in general involve inertial accelerations and the equivalence principle, which we will elaborate on, equated inertial acceleration with gravitation—inertial acceleration could not be distinguished from gravity. Thus GR also became a theory of gravitation. GR not only provided a conceptually sound understanding of gravitation, it made accurate predictions compared to Newton's law which agree with experiment.

S. Dhurandhar and S. Mitra, *General Relativity and Gravitational Waves*,
UNITEXT for Physics, https://doi.org/10.1007/978-3-030-92335-8_2

GR must inevitably be used for space based technologies essential in modern day life, e.g., global positioning system (GPS), modelling orbits of satellites etc.

2.1 Equivalence Principle: Weak Form

The principle of equivalence implies a deep connection between inertia and gravitation.

It is said that Galileo dropped massive objects from heights and observed the fall of the objects. The observation was that different objects made of different materials fell at the same rate within experimental errors and modulo other physical factors affecting the experiment. Even as of today there is no fundamental reason why this should be so. However, todays high precision experiments show within experimental error that this is indeed so. Equivalence principle has been experimentally verified to better than one part in $\sim 10^{14}$ (Touboul et al. (2017)). We make the notions more precise below. It is interesting to note that Galileo was also involved in the founding of the equivalence principle apart from the principle of relativity. In fact these are the two principles that underpin GR.

Let an object be placed in a gravitational field whose acceleration due to gravity is given by $\mathbf{g}$, then the force acting on the object is given by $\mathbf{F} = m_g\, \mathbf{g}$, where m_g is the constant of proportionality called the gravitational mass. The quantity m_g describes the response of the object to gravity. On the other hand, according to Newton's second law of motion, the acceleration of a body $\mathbf{a}$ is proportional to the force $\mathbf{F}$ applied to it, such that, $\mathbf{F} = m_i\, \mathbf{a}$, where m_i is another constant, representing the resistance of the object to motion—the inertia of the body. It is called the inertial mass of the body (the same property that appears in the definition of momentum, kinetic energy etc.). Therefore for a body falling under gravity, one can write the equation:

$$\mathbf{a} = (m_g/m_i)\, \mathbf{g}. \qquad (2.1.1)$$

Galileo's experiments indicate that the ratio m_g/m_i should be the same for all bodies. This is taken as a *principle* in GR called the *equivalence principle* in its weak form. It is called the weak equivalence principle (WEP) . The units can be chosen in such a way that the ratio $m_g/m_i = 1$. Then the weak equivalence principle asserts that $m_g = m_i$ for all bodies irrespective of their mass or other constituents. Many experiments have been performed to measure the deviation of m_g/m_i from one, but so far every experiment has produced a null result, that is, m_g/m_i has come out to be very close to unity, in conformity with the equivalence principle. This makes gravity a special kind of force. The other forces, e.g., electromagnetic forces are proportional to the charge and so produce accelerations depending on the charge to mass ratios which could be different for different objects. But gravity produces the same acceleration on every body, or in other words, the ratio of gravitational charge to its mass (inertial) is the same for all bodies. This is the key property that lends gravity to geometrisation.

It is important to remark that $m_g = m_i$ is simply a coincidence in Newton's theory. In physics, it is customary to look upon any such coincidence as suspicious. Einstein's theory of relativity (GR), far from taking it as a coincidence, takes it as a principle underpinning the theory. If it were found that this equality does not hold true, Newton's theory of gravity would still stand, but Einstein's theory would fall. This is in fact a strength of Einstein's theory.

Imagine an observer in a closed elevator. If the elevator is in a gravitational field **g**, the observer will feel a weight m_g**g**. While if the capsule was far away from every other body (negligible gravitational force) and was subject to an acceleration $-$**g**, the observer would feel a weight m_i**g**. Since $m_g = m_i$, the observer would not be able to distinguish between these two scenarios by feeling his/her weight. By a similar logic, in a freely falling elevator of small dimensions, all masses would fall at the same rate and would remain stationary with respect to each other within the small region. In this "local" region one has then got rid of gravity. It is always possible to get rid of gravity in a sufficiently small region of spacetime. What decides the size of this region? It is the non-uniformity of the gravitational field and the accuracy of the measurements that can be performed. This can be explained by a concrete example. Consider an observer who drops a ball or better two balls in a frame freely falling in the field of the earth. Then they would stay at rest with respect to each other, so the frame would appear to be inertial, at least, for a short time. However, if the balls are observed for a longer time, due to the radial and non-uniform nature of earth's gravitational field, the balls would slowly move with respect to each other—if the two ball were initially separated vertically, the separation will tend to increase, because the pull on the ball nearer to Earth would be stronger; on the other hand, if the initial separation was horizontal, the separation will tend to decrease, because the balls are falling towards the centre of the Earth—the point is that the observer will be able to measure the non-uniformity of the field by observing the relative motion of the balls. This relative motion is proportional to the separation between the balls and also the observation time. If the local region is small in space and time and hence the spatial separation and the time of observation small, the relative motion will be undetectable for a given accuracy of measurement. More accurate the measurement and more non-uniform the field, smaller should be the local region for the motion to be discernable. See Exercise 2.3.2 for detailed analysis of a local inertial frame near the Earth's surface.

The term "local" in the above paragraph is *very important*. The region both in space and time should be sufficiently small for the gravity (say measured by the acceleration it causes) to be uniform. In a non-uniform gravitational field and with non-local measurements relative motion between test particles can be detected. The cause of this relative motion is the gradient in the gravitational acceleration or differential acceleration and is called the tidal force—gravity exerts a tidal force. Yes, this is the same force that causes tides in the oceans. This is a differential force which causes elastic strain in matter and falls off as $1/r^3$ for a massive point source. This is why the moon has more effect on ocean tides, not the sun, although the sun's gravitational attraction on the earth is more than that of the moon.

2.2   Equivalence Principle: Strong Form (SEP)

The weak equivalence principle applies only to test masses moving under the influence of gravity. Einstein generalised this notion and created a stronger version of the equivalence principle. The stronger version says that the outcome of *any physical experiment* in a freely falling frame or a locally inertial frame (LIF) is the same. Thus, for example, laws of electrodynamics would be the same in every freely falling frame. Physics is the same in all LIFs—the outcome of any local physical experiment performed would be identical in each LIF. And what is this physics? This is just the physics in SR. So one may state the principle also by saying that SR is valid in every LIF .

2.3   Gravity and Curvature of Space-Time

We have now demanded that SR is valid in each LIF. But note that LIF is only local. The question is how does one describe physics in the large where locality does not hold. This of course can be achieved by stitching together the LIFs. For example for a spherically symmetric object such as the Earth, one must stitch together the freely falling LIFs which are all radially falling. This naturally leads one to the notion of a curved space or since the space and time have already been married together in SR, the notion of a curved spacetime.

Curved spaces had already been studied by Gauss and Riemann. A sphere is a very simple curved space. It can be thought of as putting small flat pieces together in just the right way. The same is true of any smooth surface—it can be built up of small flat pieces. This idea fits in very well with the above described situation. The freely falling frames or LIFs are the flat pieces in which SR is valid while fitting them together will give physics in the large. This naturally leads to the idea of a curved spacetime to describe GR or gravitation. In order to do this we will require many of the tools required to work with curved spaces which are in general known as differentiable manifolds. A surface such as a sphere is a two dimensional curved space or a differentiable manifold—two dimensional, because one needs just two coordinates, say θ, ϕ to describe the space locally. It is curved because Euclid's geometry is not valid in it. For example, if one constructs a triangle on a sphere, where now each side consists of arcs of great circles (these are the straightest possible curves that one can draw on the surface of a sphere and hence take the place of straight lines), the sum of the angles of the triangle do not add up to $180°$. But in Euclidean geometry, the angles of a triangle must add up to $180°$. This argument shows that the sphere is not a Euclidean space. It is what is called a curved space. A sphere is a curved space of two dimensions—we will define curvature later. A plane has zero curvature or it is said to be flat, while a sphere has non-zero curvature and it is said to be curved. However, in SR and GR, we need 3 coordinates for space and one coordinate in time to describe the spacetime and so we are dealing with a 4 dimensional spacetime.

In SR the spacetime is flat and indicates the absence of gravity, while in GR where there is gravity it is curved. Further, we will need tensors and tensor calculus to describe the physics—the physical fields. Since the physics in gravity necessarily involves curved spaces one must deal with general coordinate transformations—not merely Lorentz transformations between inertial frames—thus the tensors we will be dealing with must transform under these general coordinate transformations. This will be dealt with in later chapters. For now let us see what consequences we can arrive at with just the equivalence principle and freely falling frames.

2.3.1 Gravity Bends Light Rays

The strong equivalence principle creates apparent contradictions. Special relativity postulates that the velocity of light is constant and light travels in a straight line. Now imagine an observer shining a laser light in a freely falling elevator perpendicular to the motion of the elevator (of course, without knowing which way the elevator is moving). This observer should see the light ray travel in a straight line. However, if there is another observer who is stationary outside the elevator, and who experiences the gravitational field, say in the"downward" direction (here downwards is defined by the direction of the acceleration due to gravity), would see the ray bend downwards. This observer would therefore naturally conclude that light bends when in a gravitational field. This situation is depicted in Fig. 2.1. The dashed line which is straight, represents in the light ray in the local inertial frame of the falling elevator and the solid curve represents the light ray from the view point of the external stationary observer in a gravitational field.

Here we would like to mention that this is just half the story. Although the equivalence principle shows qualitatively that light must bend in a gravitational field it does not give the full answer. See Exercise 2.3.2. In fact, to delve into history, Einstein had used the equivalence principle to compute the bending of a light ray emitted

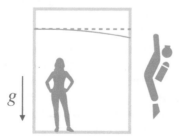

Fig. 2.1 An illustration showing that a laser light ray must bend for the outside stationary observer who experiences a gravitational field. On the other hand, for the observer falling with the elevator, since special relativity holds in an inertial frame for a freely falling observer, the light ray travels in a straight line

by a background star, just grazing the Sun's surface and observed on Earth. This was in 1911 before he had given his field equations, which he gave in 1915. The experiment to verify this prediction at that time could only be performed during a solar eclipse, when the background stars would be visible. But as it happened at that time the experiment could not be carried out because of world war I (fortunately for relativity and Einstein!). Because Einstein's computation using only the equivalence principle led to half of the observed value measured later! The reason for this discrepancy is because equivalence principle is one of the causes effecting the bending of light; there is another cause, namely, the curvature. This curvature results from the Einstein's field equations which were given in 1915. The full computation is given in Sect. 7.2.

2.3.2 *Gravity Affects Clocks: Gravitational Redshift*

In special relativity clocks run at the same rate everywhere in the spacetime in a given frame. This frame can be chosen globally throughout the spacetime. This is however not the case in general relativity. The time measurements by the clocks or the rate at which clocks run depend on the gravitational field in which they are placed and this can be different at different space-time points in a non-uniform gravitational field.

This can be argued using a thought experiment. Imagine that a laser beam is pointed vertically upwards from the earth and the photons are converted to mass at a certain height, e.g., via particle-antiparticle pair production. If these massive particles are then let to fall freely, when they reach the same height as the emission point of the laser beam, they will have gained extra kinetic energy. That is, if those particle pairs are combined now, they will have more energy than the original photons. Thus we have a perpetual motion machine that can violate conservation of energy locally.

However, this issue is resolved, if the photons lose energy as they go up, that is, their frequency reduces by a fraction equal to gh/c^2, where g is the acceleration due to gravity and h is the height of the photon with respect to the emitter. As we will see in Sect. 7.4, Einstein's theory resolves these issues in an elegant way and the effect has been experimentally tested in different ways. In fact, this has become a part of day to day life, as Global Positioning Systems (GPS), extensively used in mobile phones for navigation, must use the gravitational redshift correction, without which it would have enormously large inaccuracies [See Excercise 1 in Chap. 7].

In addition to the equivalence principle we will also require that the field equations of the theory do not change their form under coordinate transformations. General relativity assumes that fundamental physics is described by such fields. This requires the fields to be tensor fields. The algebra and the calculus of tensors is developed in subsequent chapters.

Exercises

1. Local Inertial Frame (LIF): Let us construct a LIF near the surface of the Earth. Take the Earth to be spherical with radius $R_\oplus \sim 6400$ km and mass $M_\oplus$. Then the acceleration due to gravity $g = GM_\oplus/R_\oplus^2 \sim 9.8$ metres/sec^2. Take a particle O falling under gravity near the surface of the Earth and consider another particle with a vertical separation of Δz from O. Let $\epsilon = 10^{-6}$ metres be the minimum distance measurement that O can perform. Let the particles fall for $\Delta t = 1$ second.

   (a) Find the maximum Δz so that Δz changes by $\delta z \leq \epsilon$. Show that:

$$\delta z = \frac{GM_\oplus}{R_\oplus^3} \Delta t^2 \Delta z \tag{2.3.1}$$

   Solve for Δz and hence show that $\Delta z \sim 0.64$ metres.

   (b) Now consider another particle horizontally separated from O by Δx. After time Δt, the particle's separation from O is changed by δx. Show that:

$$\delta x = -\frac{1}{2}\frac{GM_\oplus}{R_\oplus^3} \Delta t^2 \Delta x \tag{2.3.2}$$

   Solve for maximum Δx by setting $\delta x = \epsilon$ and hence show that $\Delta x \sim 1.28$ metres.

   Δx and Δz essentially give the size of the LIF. Note that the time-scale here is given by $\tau \sim \sqrt{GM_\oplus/R_\oplus^3} \sim 800$ seconds and we have carefully chosen $\Delta t \ll \tau$.

   Note that the LIF is a 4-dimensional region in spacetime.
   Thus besides extending in space it extends in time for $\Delta t = 1$ second.

   (c) Why is $\Delta x = 2\Delta z$ and $\delta x, \delta z$ opposite in sign?
   Set up a Cartesian coordinate system at O with x, y axes horizontal and z axis vertical, then at $x = y = 0$, $z = R_\oplus$ compute the second derivatives of the Newtonian potential:

$$\frac{\partial^2 \phi}{\partial x^2} = \frac{GM_\oplus}{R_\oplus^3}, \qquad \frac{\partial^2 \phi}{\partial z^2} = -2\frac{GM_\oplus}{R_\oplus^3}.$$

   Use the Laplace equation and axial symmetry about the z axis to argue your answer.
   The second partial derivatives of ϕ are essentially the components of the Riemann tensor—the tidal accelerations. The LIF can also be obtained via the geodesic deviation equation. See Chapter 5, Exercise (3).

2. The same situation as in Exercise 2.1, but now take initially a sphere of freely falling particles centred at O. All particles are at rest initially at say $t = 0$.

(a) Writing the coordinates as $x_i = \{x_1, x_2, x_3\}$, show that

$$\frac{\partial^2 \phi}{\partial x_i\, \partial x_k} = \frac{G M_\oplus}{R^3}\left(\delta_{ik} - 3\frac{x_i x_k}{R^2}\right) \equiv \phi_{,ik},$$

where δ_{ik} is the Kronecker delta. Also deduce that $\mathrm{Tr}\,[\phi_{,ik}] = 0$, which is just the Laplace equation $\nabla^2 \phi = 0$. Evaluate $\phi_{,ik}$ at $x_1 = x_2 = 0$ and $x_3 = R_\oplus$.

(b) Consider a sphere of particles falling towards the Earth near the Earth's surface. Show that the sphere deforms into an ellipsoid with the major axis along the vertical direction when it falls for a short time $\Delta t \ll \tau$. Show this as follows:

Let Δx_i be the position coordinates of a particle with respect to the reference particle O and let δx_i be the change in x_i after time Δt. Then show that,

$$\delta x_i = -\frac{1}{2}\sum_{k=1}^{3} \phi_{,ik}\,\Delta x_k\, \Delta t^2.$$

From the above equation deduce that the sphere is deformed into an ellipsoid as described by $\Delta x_i' = \Delta x_i + \delta x_i$.

(c) Show that to the first order in $\Delta t/\tau$ the ellipsoid has the same volume as the initial sphere.

3. Compute the deflection of light using only the equivalence principle for a light ray whose minimum distance from a central mass M placed at the origin is $b >> m$, where $m = GM/c^2$. In absence of the mass take the light ray travelling parallel to the x-axis with $y = b$. Let $(x(t), y(t))$ be the coordinates of the light ray and compute the bending for a light ray starting at $x = 0$ with $\dot{x} = c$, $\dot{y} = 0$.

(a) Show that:

$$\ddot{y} = -\frac{GM}{r^3}y.$$

where r is the radial coordinate.

(b) Assuming that the bending occurs only near the mass, set $y = b$, a constant along the ray, integrate the equation to obtain:

$$\dot{y} = -\frac{GM}{bc}$$

And hence compute $\delta\phi \approx \tan\delta\phi = |\dot{y}/\dot{x}| = m/b$. Using symmetry arguments, show for a light ray coming in from $x = -\infty$, the angle of deflection is $\Delta\phi = 2\delta\phi = 2m/b$.

This is just *half* the correct deflection computed from applying full GR— the other half comes from the space curvature. See Sect. 7.2 for the correct calculation.

Chapter 3
Tensor Algebra

3.1 Introduction

We have seen that gravitation is a manifestation of the curvature of spacetime. The mathematical structure which describes the physical situation had already been worked out by Gauss and Riemann several decades before Einstein conceived of GR. So the mathematical tools already existed for Einstein to use in his theory. The mathematical structure is the differentiable manifold—the spacetime of GR is a 4-dimensional differentiable manifold with the addition of a metric. Differentiable manifolds on which a metric is defined are called Riemannian manifolds. The metric is a second rank tensor. Thus one needs to understand and develop the algebra and calculus of tensors in order to proceed with GR. We will do this here step by step instead of launching directly into tensors on manifolds. We will first consider flat spaces like the spacetime of special relativity in which there is no gravity and in particular n dimensional Euclidean spaces with which the student is usually familiar. In SR we have already encountered the quantity $\eta_{\mu\nu}$ which is in fact an example of a second rank tensor but only under Lorentz transformations. However, in GR we operate with general coordinate transformations so that the tensors and in particular the metric tensor transforms under general coordinate transformations—the scope is wider. Lot of the tools such as tensor algebra and also the analysis to a large extent can be developed in flat spaces and for making a smooth transition to general curved spaces we will do so in general curvilinear coordinate systems. The advantage here is that in flat space we can always choose a Cartesian coordinate system globally pervading the entire space in which the metric coefficients are constants. We can therefore use this as an "anchor" to obtain our results. This is not so for a curved space where one must necessarily use curvilinear coordinate systems—in fact more than one of them to entirely cover the manifold. These statements will become clear later and will be validated in the text as we go along. So by the approach adopted here, we are taking one step at a time.

There is another advantage to starting with a flat space. The quantities such as vectors and tensors are elements of certain vector spaces which are associated with

S. Dhurandhar and S. Mitra, *General Relativity and Gravitational Waves*, UNITEXT for Physics, https://doi.org/10.1007/978-3-030-92335-8_3

Fig. 3.1 The shaded plane is a tangent plane to the sphere at the point P on the sphere. The tangent plane consists of vectors such as **v**. Note that the vector **v** is distinct from any point on the sphere. Credits: Ankit Bhandari

the differentiable manifold but the vectors or tensors are not elements of the manifold. The tangent space at a point of a manifold plays an important role—the (contravariant) vectors belong to this vector space. For example, consider the usual sphere which is a 2 dimensional manifold. At any point P on the sphere we can draw a tangent plane which consists of vectors tangent to the sphere's surface. These vectors are not in the sphere and the tangent plane is distinct from the sphere [see Fig. 3.1]. More importantly we cannot associate vectors on the plane in a unique way with the points on the sphere. On the other hand, the situation is different for the plane. Although the tangent space at any point of the plane is distinct from the plane (the manifold) itself, we can identify the two—to each vector in the tangent plane we can uniquely identify a point in the plane. We have the so called "position vectors". We can talk of position vectors without ambiguity and use them to our advantage. Tensors are elements of vector spaces built up essentially from the tangent space and its dual (vector space of linear mappings from the tangent space to the real numbers—this is also a vector space with the same dimension as the tangent space and is called the cotangent space). The tangent space has the same dimension as the manifold. A differentiable manifold is locally flat. For example a small region of the sphere looks like a plane. This is the reason why ancient people thought that the earth was flat—a region the size of few kilometres looks flat although the earth is approximately a sphere. For this purpose the size of the region must be small compared with the radius of the earth. Generally, at each point of the manifold, we can construct a tangent space because the manifold is locally flat. This is the key reason why we will be able to port the results of tensor algebra obtained for a flat space on to a curved manifold. At this point there is no loss of generality to our approach.

In this book we will not enter into the formal discussion of modern mathematics mentioned in the previous paragraph, but we will take a toned down and intuitive approach which will essentially give sufficient understanding of the subject, and further deduce and appreciate the results obtained in this book.

3.2 Basis Vectors

First we will describe a n dimensional flat space. Any point in this space can be specified by n real numbers called its coordinates denoted by x^i, $i = 1, 2, ..., n$, where we have used a superscript to label the coordinates. Note that it is not an exponent and so any such confusion should be avoided. Strictly speaking a curvilinear coordinate system that is not Cartesian does not cover the entire manifold but a Cartesian system does. At this point, we will call a space flat if it can be covered fully by x^i and the distance between adjacent points labelled by x^i and $x^i + dx^i$ is given by:

$$ds^2 = \eta_{ik} dx^i dx^k \,, \tag{3.2.1}$$

where η_{ik} is the metric tensor which is symmetric, namely, $\eta_{ik} = \eta_{ki}$ and the η_{ik} are *constants*. This we will see is fully equivalent to the description in Chapter 4 of a flat space in terms of the Riemann tensor, in which the Riemann tensor vanishes identically throughout the space (see Exercise 4.3 in Chapter 4). By a principal axis transformation and rescaling, we can make the η_{ik} diagonal and each diagonal entry equals ± 1. The spacetime of SR is an example of a 4 dimensional flat space. If the diagonal entries are all $+1$, then ds^2 is a sum of squares which is in fact the repeated application of the Pythagoras theorem. Such a space is called Euclidean and the metric is called positive definite. The SR spacetime has an indefinite metric because here the terms in the metric η_{ik}, in the convention we have adopted, there is one positive sign and three negative signs. We will in general consider any curvilinear coordinate system which may not necessarily span the full space. This does not hamper us in any way because we will be only doing local analysis. As remarked before, in flat space it is possible to specify a point $\mathbf{x}$ uniquely, namely the position vector, using n coordinates $x^i, i = 1, ..., n$. One can construct "imaginary grid lines" for mental visualisation of the space coordinates. For a given k, along the x^k grid line only the coordinate x^k varies, while others are kept constant. One can also imagine $n - 1$ dimensional hypersurfaces which slice the space, such that on each slice only one coordinate is held constant. Surfaces for which a given coordinate is kept constant should not intersect, as that will make a point in space to have two values of the same coordinate, hence not a valid coordinate system.

In order to make the discussion simple and familiar we will consider a positive definite metric. The results are virtually similar and easily generalised to an indefinite metric. To represent a vector field $\mathbf{V}(\mathbf{x})$ in this space in terms of its components we need to define n basis vectors at each point in space. Let $\mathbf{x}$ be the position vector depending on n curvilinear coordinates x^i, that is, $\mathbf{x}(x^i)$. The x^i form a grid of coordinate curves covering the space. The basis vectors can be chosen in two natural ways:

1. Along the grid lines: if the ith coordinate is changed from x^i to $x^i + dx^i$, the direction in which the position vector $\mathbf{x}$ changes (which is along the grid lines) could be taken as the direction of the ith basis vector $\mathbf{e}_i$ at $\mathbf{x}$. That is,

$$\mathbf{e}_i := \frac{\partial \mathbf{x}}{\partial x^i} \,. \tag{3.2.2}$$

   Therefore geometrically, each $\mathbf{e}_i$ is a tangent to the coordinate curve x^i.

2. Normal to the slicing surfaces: the gradient to the surface of constant x^i at $\mathbf{x}$ is normal to the surface. Here the basis vectors would be

$$\mathbf{e}^i := \nabla x^i \,, \tag{3.2.3}$$

   where the gradient of a function can be defined as a vector whose magnitude is the rate of change of the function and whose direction is one for which the function varies fastest. Thus each $\mathbf{e}^i$ is normal to the surface $x^i = $ const.

Note that each basis vector is a vector field and defined at each point in the region where the coordinates are defined. The n basis vector fields form a basis at each point. Both $\mathbf{e}_i$ and $\mathbf{e}^i$ are functions of $\mathbf{x}$. This is called a *coordinate* basis. The basis vectors are not necessarily normalised. Generally we are used to normalised "unit" vectors as basis vectors, though there is no loss of generality if one uses unnormalized basis vectors. A vector field $\mathbf{V}(\mathbf{x})$ can then be expanded in terms of the basis vectors at any given point $\mathbf{x} = \mathbf{x}_0$ in either of these bases taken at $\mathbf{x}_0$ as:

$$\mathbf{V} = \sum_{i=1}^{n} \mathbf{e}_i \, V^i = \sum_{i=1}^{n} \mathbf{e}^i \, V_i \,. \tag{3.2.4}$$

Since we are at a given point $\mathbf{x}_0$, we leave out the position dependence explicitly. But the above equation is true at each point of the defined region. Note also the superscript and subscript on the components of the vector $\mathbf{V}$. We will discuss this point in the next section. In an orthogonal curvilinear coordinate system, the slicing surfaces would meet at right angles and one could show that the above two sets of basis vectors become co-aligned ($\mathbf{e}_i \propto \mathbf{e}^i$). However, in general this is not true. In a general curvilinear coordinate, the orthogonality of the basis vectors does not hold. However, the basis vectors, $\mathbf{e}^i$ and $\mathbf{e}_i$ defined in the above manner, form a reciprocal system of vectors, such that,

$$\mathbf{e}^i \cdot \mathbf{e}_j = \delta^i_j \,, \tag{3.2.5}$$

where

$$\delta^i_j := \begin{cases} 1 \text{ if } i = j \\ 0 \text{ otherwise} \end{cases}, \tag{3.2.6}$$

is the Kronecker Delta function. The concept of a non-orthogonal curvilinear coordinate system and how the basis vectors form a reciprocal system of vectors is made clear with the specific example of a two-dimensional oblique coordinate system on a plane in Sect. 3.7. Note, however, that this example is for a coordinate system, that can cover the whole manifold, which is simpler to understand. In general, a single coordinate system may not even suffice to cover the whole space.

The above equations can be used to extract the components along a given unit vector. Note that we can talk of unit vectors and dot products because we are in Euclidean space on which the usual metric is defined. The ith component of $\mathbf{V}$ along the basis vector $\mathbf{e}_i$ and $\mathbf{e}^i$ can thus be obtained *by projecting the vector on the reciprocal vector*.

$$V^i = \mathbf{V} \cdot \mathbf{e}^i, \tag{3.2.7}$$
$$V_i = \mathbf{V} \cdot \mathbf{e}_i. \tag{3.2.8}$$

Again, this is very different from the special case of orthogonal coordinates.

Derivation 1 *Show that the two sets of basis vectors $\mathbf{e}_i$ and $\mathbf{e}^i$ form a reciprocal system of vectors (Spiegel (1959)).*

Using the chain rule of partial differentiation one can write

$$d\mathbf{x} = \sum_{i=1}^{n} \frac{\partial \mathbf{x}}{\partial x^i} dx^i = \sum_{i=1}^{n} \mathbf{e}_i \, dx^i. \tag{3.2.9}$$

Taking dot product of each side with $\mathbf{e}^j := \nabla x^j$, one gets

$$\nabla x^j \cdot d\mathbf{x} := dx^j = \sum_{i=1}^{n} (\mathbf{e}_i \cdot \mathbf{e}^j) \, dx^i. \tag{3.2.10}$$

Since the above identity has to hold for arbitrary $d\mathbf{x}$, comparing the coefficients of dx^i on both sides we obtain $\mathbf{e}_i \cdot \mathbf{e}^j = \delta^j_i$.

3.3 Contravariant and Covariant Components of a Vector

About the notation, we have used the same case for both kinds of components as that of the vector field. The V^i are called "contravariant" components of a vector.[1] Similarly, V_i are called "covariant" components of a vector. In fact, we follow a

---

[1] We generally omit the phrase "components of", as the meaning is clear from the context even without that phrase.

similar convention even for the basis vectors but as observed it is opposite to the components of a vector.

Let us now consider coordinate transformations. Consider another coordinate system x'^i. In the common region where x^i and x'^i are both defined and x'^i are functions of x^j, that is $x'^i(x^j)$ and vice-versa. The only condition is that the Jacobian matrix of the transformation $\partial x'^i/\partial x^j$ should be non-singular so that the transformations are locally invertible - they form a valid coordinate system. Contravariant and covariant components of vectors and corresponding basis vectors transform differently under coordinate transformations.

Derivation 2 *Find how the basis vectors change under the coordinate transformation $x^i \to x'^i = x'^i(x^j)$.*

Coordinate transformations change the components of a vector, but not the vector itself. Hence in the primed coordinates

$$\mathbf{e}'_i := \frac{\partial \mathbf{x}}{\partial x'^i} = \sum_{j=1}^{n} \frac{\partial \mathbf{x}}{\partial x^j} \frac{\partial x^j}{\partial x'^i} = \sum_{j=1}^{n} \frac{\partial x^j}{\partial x'^i} \mathbf{e}_j . \tag{3.3.1}$$

To find how the other set of basis vectors, $\mathbf{e}^i := \nabla x^i$, transform we notice that

$$\mathbf{e}'^i \cdot \mathbf{dx} = \mathrm{d}x'^i = \sum_{j=1}^{n} \frac{\partial x'^i}{\partial x^j} \mathrm{d}x^j = \sum_{j=1}^{n} \frac{\partial x'^i}{\partial x^j} \mathbf{e}^j \cdot \mathbf{dx} . \tag{3.3.2}$$

If this identity has to hold for any arbitrary $\mathbf{dx}$, one should have

$$\mathbf{e}'^i = \sum_{j=1}^{n} \frac{\partial x'^i}{\partial x^j} \mathbf{e}^j ; \quad \mathbf{e}'^i \cdot \mathbf{e}_j = \frac{\partial x'^i}{\partial x^j} . \tag{3.3.3}$$

Derivation 3 *Find how the components of a vector* **V** *transform for the change of coordinate* $x^i \rightarrow x'^i = x'^i(x^j)$.

Since a vector does not change in the new coordinates, only its components do,

$$\mathbf{V} = \sum_{i=1}^{n} V^i \, \mathbf{e}_i = \sum_{i=1}^{n} V_i \, \mathbf{e}^i = \sum_{i=1}^{n} V'^i \, \mathbf{e}'_i = \sum_{i=1}^{n} V'_i \, \mathbf{e}'^i . \tag{3.3.4}$$

Then using eqn. (3.2.7, 3.2.8, 3.2.5, 3.3.1 & 3.3.3) one can write

$$V'^i = \mathbf{V} \cdot \mathbf{e}'^i = \sum_{k=1}^{n} V^k \, \mathbf{e}_k \cdot \sum_{j=1}^{n} \frac{\partial x'^i}{\partial x^j} \, \mathbf{e}^j = \sum_{j=1}^{n} \frac{\partial x'^i}{\partial x^j} \, V^j \tag{3.3.5}$$

$$V'_i = \mathbf{V} \cdot \mathbf{e}'_i = \sum_{k=1}^{n} V_k \, \mathbf{e}^k \cdot \sum_{j=1}^{n} \frac{\partial x^j}{\partial x'^i} \, \mathbf{e}_j = \sum_{j=1}^{n} \frac{\partial x^j}{\partial x'^i} \, V_j . \tag{3.3.6}$$

If we take $\mathbf{e}_i$ as the basis, then we see that the covariant components transform in the same way as this basis while the contravariant components transform with the inverse matrix. This is the historical reason for this nomenclature.

Physical quantities for one, do not change under coordinate transformations, but components of vectors and tensors do. That is why physical quantities appear as pairs of covariant and contravariant components summed over the same index, like in Eq. (3.3.4), because this procedure constructs coordinate independent quantities. This motivates us to consider the summation signs, in, e.g., Eq. (3.3.4), redundant. Therefore, whenever we see the same index twice in an expression, one in the subscript and one in the superscript, a summation over all allowed values of the index (normally 1 to n or 0 to 3 for space-time coordinates) is implied, unless explicitly specified otherwise. This is called *Einstein's summation convention*, a very useful shorthand, which some people jokingly say is Einstein's only contribution to mathematics!

For example, using the summation convention one can succinctly write the transformation properties of the covariant and contravariant components as,

$$V'^i = \frac{\partial x'^i}{\partial x^j} \, V^j ; \quad V'_i = \frac{\partial x^j}{\partial x'^i} \, V_j . \tag{3.3.7}$$

The summation convention is used in the rest of this book, unless, otherwise specified.

3.3.1 The Jacobian matrices

The quantities entering Eq. (3.3.7), namely, $\partial x'^i / \partial x^j$ and $\partial x^j / \partial x'^i$ are special matrices called Jacobian matrices. When coordinates are transformed, the new coordinates should be a valid set of coordinates that uniquely specify a point in the n-dimensional space. Let the old system of coordinates be x^i and the new coordinate system be x'^k. Then x'^k is a valid coordinate system if and only if, the Jacobian matrix:

$$
\mathbf{J} = \left\| \frac{\partial x'^i}{\partial x^j} \right\| \equiv \frac{\partial(x'^1, x'^2, \cdots, x'^m)}{\partial(x^1, x^2, \cdots, x^n)} := \left\|
\begin{array}{cccc}
\dfrac{\partial x'^1}{\partial x^1} & \dfrac{\partial x'^1}{\partial x^2} & \cdots & \dfrac{\partial x'^1}{\partial x^n} \\
\dfrac{\partial x'^2}{\partial x^1} & \dfrac{\partial x'^2}{\partial x^2} & \cdots & \dfrac{\partial x'^2}{\partial x^n} \\
\vdots & \vdots & \vdots & \vdots \\
\dfrac{\partial x'^m}{\partial x^1} & \dfrac{\partial x'^m}{\partial x^2} & \cdots & \dfrac{\partial x'^m}{\partial x^n}
\end{array}
\right\|, \tag{3.3.8}
$$

is non-singular. This is true if and only if its determinant, which we denote by J, does not vanish. The vanishing of the determinant means that not all the x'^i are independent functions of x^i, that is, there are not n independent transformed coordinates. Here we have invoked the inverse function theorem: The functions $x'^i(x^k)$ are locally invertible if and only if the $J = \det(\partial x'^i/\partial x^k) \neq 0$. See Exercise 3.3 for a 2 dimensional example when this condition does not hold.

The same applies to the inverse transformations when x^i are considered as functions of x'^j. Then the relevant Jacobian matrix is:

$$
\mathbf{J}' = \left\| \frac{\partial x^i}{\partial x'^j} \right\| \equiv \frac{\partial(x^1, x^2, \cdots, x^n)}{\partial(x'^1, x'^2, \cdots, x'^m)} := \left\|
\begin{array}{cccc}
\dfrac{\partial x^1}{\partial x'^1} & \dfrac{\partial x^1}{\partial x'^2} & \cdots & \dfrac{\partial x^1}{\partial x'^m} \\
\dfrac{\partial x^2}{\partial x'^1} & \dfrac{\partial x^2}{\partial x'^2} & \cdots & \dfrac{\partial x^2}{\partial x'^m} \\
\vdots & \vdots & \vdots & \vdots \\
\dfrac{\partial x^n}{\partial x'^1} & \dfrac{\partial x^n}{\partial x'^2} & \cdots & \dfrac{\partial x^n}{\partial x'^m}
\end{array}
\right\|, \tag{3.3.9}
$$

In fact the matrices $\mathbf{J}$ and $\mathbf{J}'$ are inverses of each other,. We have $\mathbf{J} \cdot \mathbf{J}' = I_n$, where I_n is the $n \times n$ identity matrix. Consequently, their determinants are inverses of each other, that is, $J' = 1/J$.

The J and J' play an important role in defining the proper volume element and integration. We discuss these issues in Sect. 3.6.

3.4 The Metric

3.4.1 Scalar Product

The scalar product of two vector fields $\mathbf{u}$ and $\mathbf{v}$ can be written in three ways:

$$\mathbf{u} \cdot \mathbf{v} = \begin{cases} (u^i\, \mathbf{e}_i) \cdot (v_j\, \mathbf{e}^j) = (u_i\, \mathbf{e}^i) \cdot (v^j\, \mathbf{e}_j) = u^i\, v_i = u_i\, v^i \\ (u^i\, \mathbf{e}_i) \cdot (v^j\, \mathbf{e}_j) = g_{ij}\, u^i\, v^j \\ (u_i\, \mathbf{e}^i) \cdot (v_j\, \mathbf{e}^j) = g^{ij}\, u_i\, v_j \end{cases} . \tag{3.4.1}$$

The first expression shows that the scalar product is the sum of product of pairs of contravariant and covariant components (that is, components in two different basis) with the same index. Which is slightly counter intuitive, as in orthogonal coordinates one writes $\mathbf{u} \cdot \mathbf{v} = \Sigma_{i=1}^{n} u_i\, v_i$ (sum of product of components in the same basis).

In the second expression, we have defined a two indexed quantity called the "metric" $g_{ij} := \mathbf{e}_i \cdot \mathbf{e}_j$, which connects the scalar product with contravariant components in the same basis. The third expression does the same with the corresponding contravariant metric tensor $g^{ij} := \mathbf{e}^i \cdot \mathbf{e}^j$. It is easy to show that these two index quantities are inverse to each other.

$$g^{ij}\, g_{jk} = (\mathbf{e}^i \cdot \mathbf{e}^j)(\mathbf{e}_j \cdot \mathbf{e}_k) = \mathbf{e}^i \cdot [\mathbf{e}^j (\mathbf{e}_j \cdot \mathbf{e}_k)] = \mathbf{e}^i \cdot \mathbf{e}_k = \delta_k^i . \tag{3.4.2}$$

The third step follows from the fact that any vector, $\mathbf{V}$, can be expressed as, $\mathbf{V} = v_j\, \mathbf{e}^j = (\mathbf{V} \cdot \mathbf{e}_j)\, \mathbf{e}^j$.

Thus, the square of an interval ds^2 can also be written in three ways,

$$ds^2 = dx^i\, dx_i = g_{ij}\, dx^i\, dx^j = g^{ij}\, dx_i\, dx_j . \tag{3.4.3}$$

3.4.2 Raising and Lowering of Indices

A contravariant component can be expressed in terms of covariant components of the vector field $\mathbf{v}$ and vice versa using Eq. (3.2.7, 3.2.8 and 3.3.4) as follows:

$$v_i = \mathbf{v} \cdot \mathbf{e}_i = v^j\, \mathbf{e}_j \cdot \mathbf{e}_i = g_{ij}\, v^j , \tag{3.4.4}$$

$$v^i = \mathbf{v} \cdot \mathbf{e}^i = v_j\, \mathbf{e}^j \cdot \mathbf{e}^i = g^{ij}\, v_j . \tag{3.4.5}$$

Thus, the metric tensors can be used for raising and lowering of indices, that is, to transform from contravariant to covariant and vice versa.

3.5 Tensors of Higher Rank

3.5.1 Transformation Laws

From this point onwards, we will not write the basis vectors $\mathbf{e}_i$ or $\mathbf{e}^i$ explicitly. Just the contravariant and covariant components of vectors/tensors and the metric would take care of the algebra. That is why we often drop or abuse the phrase "components of" and use contravariant or convariant vector/tensor to mean that they are components of those vectors/tensors. There are also mixed tensors with both covariant and contravariant indices, e.g., the Kronecker Delta symbol δ^i_j.

Components of tensors transform following similar rules as those of vectors. We will first consider second rank tensors and indicate how the transformation law generalises to tensors of higher rank. Under a coordinate transformation $x^i \rightarrow x'^i = x'^i(x^j)$, components of a contravariant tensor T^{ij}, covariant tensor T_{ij} and a mixed tensor T^i_j transform respectively as:

$$T'^{kl} = T^{ij} \frac{\partial x'^k}{\partial x^i} \frac{\partial x'^l}{\partial x^j}, \tag{3.5.1}$$

$$T'_{kl} = T_{ij} \frac{\partial x^i}{\partial x'^k} \frac{\partial x^j}{\partial x'^l}, \tag{3.5.2}$$

$$T'^k_l = T^i_j \frac{\partial x'^k}{\partial x^i} \frac{\partial x^j}{\partial x'^l}. \tag{3.5.3}$$

The pattern is clear. If a tensor has more contravariant indices then more matrices of the form Eq. (3.3.8) enter into the transformation law. In case of covariant indices the inverse of the matrix given in Eq. (3.3.9) enters. Thus one may write down the tensor transformation law for a tensor with m contravariant indices and n covariant indices. Such a tensor is of rank $m + n$.

We give below two examples of tensors of rank 2, contravariant, covariant and mixed which we have encountered before but perhaps not recognised them as such.

Derivation 4 *Show that g_{ij} and g^{ij} transform respectively as covariant and contravariant tensors under coordinate transformation $x^i \rightarrow x'^i = x'^i(x^i)$*

Using the results obtained in Derivation 2, it is straightforward to show that

$$g'_{ij} = \mathbf{e}'_i \cdot \mathbf{e}'_j = \left(\frac{\partial x^i}{\partial x'^k} \mathbf{e}_k \right) \cdot \left(\frac{\partial x^j}{\partial x'^l} \mathbf{e}^l \right) = \frac{\partial x^i}{\partial x'^k} \frac{\partial x^j}{\partial x'^l} g_{kl}, \tag{3.5.4}$$

$$g'^{ij} = \mathbf{e}'^i \cdot \mathbf{e}'^j = \left(\frac{\partial x'^i}{\partial x^k} \mathbf{e}_k \right) \cdot \left(\frac{\partial x'^j}{\partial x^l} \mathbf{e}^l \right) = \frac{\partial x'^i}{\partial x^k} \frac{\partial x'^j}{\partial x^l} g^{kl}. \tag{3.5.5}$$

Another way is to use the quotient law. But then the argument is somewhat involved. We may start from Eq. (3.4.3) that since ds^2 is a scalar and dx^i is an arbitrary contravariant vector then g_{ij} must be a covariant tensor of rank 2. Not true! Although dx^i is an arbitrary contravariant vector $dx^i dx^j$ is not arbitrary because it is symmetric. This argument will only show that the symmetric part of g_{ij} is a tensor. See Exercise 3.5b. However, if in addition, we impose the symmetry on g_{ij} then the proof is complete. A similar reasoning shows that g^{ij} must be a contravariant tensor of rank 2.

> **Derivation 5** *Show that Kronecker Delta δ_j^i transform as a mixed tensor.*
> *Since the transformed δ_j^i symbols match $\delta_j^{\prime i}$ in the new coordinates,*
>
> $$\delta_j^i \frac{\partial x'^k}{\partial x^i} \frac{\partial x^j}{\partial x'^l} = \frac{\partial x'^k}{\partial x'^l} = \delta_l^{\prime k}. \tag{3.5.6}$$
>
> *the proof follows.*

A theorem, which is often useful is the **Quotient Rule**. There are two of them: one for the outer product and one for the inner product. First consider the outer product of tensors A_{ij} and B^k say, $C_{ij}^k = A_{ij}B^k$. Then if C_{ij}^k and B^k are tensors, then A_{ij} is a tensor. We have taken here a simple case but this result is true for tensors of arbitrary rank.

Now consider the inner product $C_i = A_{ij}B^j$, where we are given that C_i and B^j are tensors. Then A_{ij} is a tensor if this equation holds for *arbitrary B^j* (Exercise 3.5). Again this statement is true for tensors of arbitrary rank.

3.5.2 Symmetric and Anti-symmetric Tensors

A tensor $T_{...i...j...}$ is symmetric in the indices i and j, if

$$T_{...j...i...} = T_{...i...j...}, \tag{3.5.7}$$

and antisymmetric if

$$T_{...j...i...} = -T_{...i...j...}. \tag{3.5.8}$$

The idea of symmetry or antisymmetry can be extended to several indices for tensors of higher rank. In the extreme case of all indices, the tensor is totally symmetric if the components do not change for any permutation of indices. It is totally antisymmetric if they only change sign under an odd permutation of the indices and remain same for even permutation.

It is always possible to decompose a tensor T_{ij} in a symmetric part, $T_{(ij)}$, and an anti-symmetric part $T_{[ij]}$. The following equations make this point clear.

$$T_{ij} = T_{(ij)} + T_{[ij]}, \tag{3.5.9}$$

$$T_{(ij)} = \frac{1}{2}(T_{ij} + T_{ji}), \tag{3.5.10}$$

$$T_{[ij]} = \frac{1}{2}(T_{ij} - T_{ji}). \tag{3.5.11}$$

The idea of symmetrisation and antisymmetrisation can be extended to tensors of higher rank. The following example for tensors of rank 3 makes this clear,

$$T_{(ijk)} = \frac{1}{3!}(T_{ijk} + T_{jki} + T_{kij} + T_{ikj} + T_{jik} + T_{kji}), \quad (3.5.12)$$

$$T_{[ijk]} = \frac{1}{3!}(T_{ijk} + T_{jki} + T_{kij} - T_{ikj} - T_{jik} - T_{kji}). \quad (3.5.13)$$

3.6 The Volume Element

3.6.1 Determinant of the Metric Tensor

Determinant of the covariant metric tensor g_{ij}, denoted by g, is a very useful quantity. Since g^{ij} is the inverse matrix of g_{ij}, the determinant of g^{ij} is $1/g$.

Since the metric tensor transforms as

$$g'_{kl} = \frac{\partial x^i}{\partial x'^k} \frac{\partial x^j}{\partial x'^l} g_{ij}, \quad (3.6.1)$$

and, determinant of product of matrices is the product of determinant of matrices, by taking determinant of both sides one gets

$$g' = J'^2 g. \quad (3.6.2)$$

Here the Jacobian $J' = \det(\partial x^i/\partial x'^k)$ which is the determinant of the inverse Jacobian matrix in Eq. (3.3.9).

Quantities which transform as tensors under coordinate transformation, except for a factor which is a function of the Jacobian, are called tensor densities and the power of the Jacobian in the factor is called the weight of the density (A. K. Raychaudhuri (1992)). Thus, g transforms as a scalar density of weight $+2$.

The differential of the determinant of the metric, dg, is also a frequently used quantity. In the differential of g, the coefficient of a given element dg_{ij}, is the cofactor G^{ij} of g_{ij} in the determinant g. However, the inverse matrix is $g^{ij} = G^{ij}/g$. Therefore we obtain,

$$dg = G^{ij} dg_{ij} = g\, g^{ij} dg_{ij} = -g\, g_{ij} dg^{ij}. \quad (3.6.3)$$

The last step follows from the fact that $d(g^{ij} g_{jk}) = d\delta_k^i = 0$. This result could also be derived starting from the differential of the determinant of g^{ij}, where dg would have to be replaced by $d(1/g) = -dg/g^2$.

The above equation can be cast into other useful forms,

$$g^{ij} dg_{ij} = \frac{dg}{g} = d\ln g; \quad \frac{1}{2} g^{ij} dg_{ij} = \frac{1}{2} \frac{dg}{g} = d\ln\sqrt{|g|} = \frac{d\sqrt{|g|}}{\sqrt{|g|}}. \quad (3.6.4)$$

3.6.2 Perfectly Anti-Symmetric Tensor: Levi-Civita Symbol

The Levi-Civita symbol,

$$\epsilon^{ijk\cdots} := \begin{cases} 0 \text{ if any two indices are equal} \\ 1 \text{ if the indices can be arranged in ascending order,} \\ \quad (1, 2, \cdots), \text{ with } even \text{ number of exchanges} \\ -1 \text{ if the indices can be arranged in ascending order,} \\ \quad (1, 2, \cdots), \text{ with } odd \text{ number of exchanges} \end{cases} \quad (3.6.5)$$

is fully antisymmetric in its indices.

A very important use of this symbol comes in the definition of determinant of a matrix. One can show that, the determinant of a matrix $\mathbf{A} \equiv A^i{}_j$ (not necessarily a tensor) can be written as

$$\det \mathbf{A} \equiv \det A^i{}_j = \epsilon^{ijk\cdots} A^1{}_i A^2{}_j A^3{}_k \cdots, \quad (3.6.6)$$

which is equivalent to the following relation,

$$\epsilon^{i'j'k'\cdots} \det \mathbf{A} = \epsilon^{ijk\cdots} A^{i'}{}_i A^{j'}{}_j A^{k'}{}_k \cdots. \quad (3.6.7)$$

We can use the above relation to find the transformation properties of these symbols. Coordinate transforming the Levi-Civita symbols and using the above relation, where $A^i{}_j = \partial x'^i/\partial x^j$, one gets

$$\frac{\partial x'^{i'}}{\partial x^i} \frac{\partial x'^{j'}}{\partial x^j} \frac{\partial x'^{k'}}{\partial x^k} \cdots \epsilon^{ijk\cdots} = \epsilon^{i'j'k'\cdots} J, \quad (3.6.8)$$

where J is the Jacobian for the transformation from unprimed to primed coordinates. Since $(\partial x'^i/\partial x^j)(\partial x^j/\partial x'^k) = \delta^i_k$, one gets, $J' = 1/J$. Thus, $\epsilon^{ijk\cdots}$ also transforms as a tensor density of weight $+1$. Hence, one can define the fully antisymmetric contravariant tensor in curvilinear coordinates as,

$$\varepsilon^{ijk\cdots} := \epsilon^{ijk\cdots}/\sqrt{|g|}. \quad (3.6.9)$$

The corresponding covariant tensor, $\varepsilon_{ijk\cdots}$, can be obtained by lowering the indices with the metric tensor in the standard way.

3.6.3 Volume Element

In three dimensional coordinates, the volume of the parallelepiped with sides $\mathbf{a}$, $\mathbf{b}$ and $\mathbf{c}$ is given by the triple product $V = \mathbf{a} \cdot (\mathbf{b} \times \mathbf{c})$, which can also be written as the determinant of the matrix constructed by the vectors and their elements

$$V = \begin{vmatrix} a_1 & a_2 & a_3 \\ b_1 & b_2 & b_3 \\ c_1 & c_2 & c_3 \end{vmatrix}.$$ (3.6.10)

This definition can be extended in a natural way to n-dimension as,

$$V = \begin{vmatrix} a_1 & a_2 & a_3 & \cdots \\ b_1 & b_2 & b_3 & \cdots \\ c_1 & c_2 & c_3 & \cdots \\ \vdots & \vdots & \vdots & \ddots \end{vmatrix}.$$ (3.6.11)

In this section, we will use upper case indices (e.g., I) when no summation is to be carried out over the index. When there is summation over such an index we will use a lowercase index following summation convention.

An infinitesimal volume element in an n-dimensional curvilinear coordinates can be defined as the parallelepiped defined by the n infinitesimal vectors $(\partial \mathbf{x}/\partial x^I)\,dx^I = \mathbf{e}_I\,dx^I$ (no summation over I, as mentioned above). We follow the same procedure as above: we first find the components of those vectors in a *local orthogonal Cartesian coordinate system* defined by $\mathbf{e}'_i$, such that, $\mathbf{e}'^1 = \mathbf{e}'_1 = \hat{\mathbf{i}}, \mathbf{e}'^2 = \mathbf{e}'_2 = \hat{\mathbf{j}}, \mathbf{e}'^3 = \mathbf{e}'_3 = \hat{\mathbf{k}}, \cdots$, arrange them in columns and take determinant. To do so, we define the matrix $A^{i'}_{\ I} := \mathbf{e}'^{i'} \cdot \mathbf{e}_I\,dx^I$. Then we use the formula

$$dV := \det \mathbf{A} = \epsilon^{ijk\cdots} A^1_{\ i} A^2_{\ j} A^3_{\ k} \cdots,$$ (3.6.12)

which can then be rewritten as

$$\epsilon^{i'j'k'\cdots}\,dV = \epsilon^{ijk\cdots} A^{i'}_{\ i} A^{j'}_{\ j} A^{k'}_{\ k} \cdots = \epsilon^{ijk\cdots} (\mathbf{e}'^{i'} \cdot \mathbf{e}_i)\,dx^i\,(\mathbf{e}'^{j'} \cdot \mathbf{e}_j)\,dx^j\,(\mathbf{e}'^{k'} \cdot \mathbf{e}_k)\,dx^k \ldots .$$ (3.6.13)

Because $\epsilon^{ijk\cdots}$ takes a nonzero value only when all the indices have different values, terms that survive in the above summation will have the common factor $dx^1\,dx^2\,dx^3 \ldots$. Hence, one can write,

$$\epsilon^{i'j'k'\cdots}\,dV = \epsilon^{ijk\cdots} (\mathbf{e}'^{i'} \cdot \mathbf{e}_i)\,(\mathbf{e}'^{j'} \cdot \mathbf{e}_j)\,(\mathbf{e}'^{k'} \cdot \mathbf{e}_k) \cdots dx^1\,dx^1\,dx^3 \ldots .$$ (3.6.14)

However, getting the components of a vector in local Cartesian coordinates is essentially a coordinate transformation and we recall that,

$$\mathbf{e}'^{i'} \cdot \mathbf{e}_i = \frac{\partial x'^{i'}}{\partial x^i}.$$ (3.6.15)

This immediately gives,

$$\epsilon^{i'j'k'\cdots} \, dV = \epsilon^{ijk\cdots} \frac{\partial x'^{i'}}{\partial x^i} \frac{\partial x'^{j'}}{\partial x^j} \frac{\partial x'^{k'}}{\partial x^k} \cdots \, dx^1 \, dx^1 \, dx^3 \cdots, \quad (3.6.16)$$

$$= \epsilon^{i'j'k'\cdots} \, J \, dx^1 \, dx^1 \, dx^3 \cdots. \quad (3.6.17)$$

Since in Cartesian coordinates $\sqrt{|g'|} = 1$, $J = 1/J' = \sqrt{|g|}$. Then setting $i' = 1, j' = 2, k' = 3, \cdots$, so that $\epsilon^{i'j'k'\cdots} = 1$, we arrive at the formula for the "proper" volume element

$$dV = \sqrt{|g|} \, dx^1 \, dx^2 \, dx^3 \cdots. \quad (3.6.18)$$

This volume element is of course invariant, because it is the volume of the same small region of the space in any coordinate system. To convince ourselves, we calculate the n-dimensional volume element in a different coordinate system:

$$dV' := \sqrt{|g'|} \, dx'^1 \, dx'^2 \, dx'^3 \cdots \equiv \sum_{i',j',\cdots} (\epsilon^{i'j'k'\cdots}/n!) \sqrt{|g'|} \, dx'^{i'} \, dx'^{j'} \, dx'^{k'} \cdots$$

$$= \sqrt{|g'|} \, (\epsilon^{i'j'k'\cdots}/n!) \frac{\partial x'^{i'}}{\partial x^i} \frac{\partial x'^{j'}}{\partial x^j} \frac{\partial x'^{k'}}{\partial x^k} \cdots \, dx^i \, dx^j \, dx^k \cdots$$

$$= \sqrt{|g'|} \, J' \sum_{i,j,\cdots} (\epsilon^{ijk\cdots}/n!) \, dx^i \, dx^j \, dx^k \cdots = \sqrt{|g|} \, dx^1 \, dx^2 \, dx^3 \cdots = dV.$$

This shows that dV is a scalar.

3.7 Example: Oblique Coordinate System on a Plane

We will work out the formal algebra presented above for a two-dimensional toy case. We construct a coordinate system, where the constant "X" and "Y" lines are not orthogonal, but they are at an angle ω, as shown in Fig. 3.2. This example has also been described in Narlikar (2010), however here we give more details of the calculation.

Utilising the fact that vectors do not change under coordinate transformations, only their components do, we will compute all the quantities in an orthogonal Cartesian coordinate system. We use the lower case letters, x and y, to represent this coordinate system. The x-axis is aligned with the $X = 0$ line and the y-axis is orthogonal to the x-axis maintaining the right-hand rule at the common origin O. In Fig. 3.2, the x-axis (and the unit vector $\hat{\mathbf{i}}$) is coincident with the $\mathbf{e}_X$ direction and the y-axis (and the unit vector $\hat{\mathbf{j}}$) is coincident with the $\mathbf{e}^Y$ direction (reason for this overlap of axes will become clear in a moment).

An arbitrary point P, which has the Cartesian coordinates (x, y), will have coordinates (X, Y) in the oblique coordinates following the relation

Fig. 3.2 A two dimensional
toy problem with tilted axes
to illustrate Illustration of
curvilinear coordinates

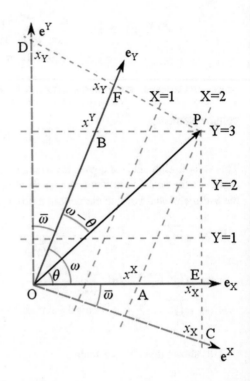

$$X = x - y \cot \omega. \tag{3.7.1}$$
$$Y = y / \sin \omega. \tag{3.7.2}$$

3.7.1 Contravariant and Covariant Bases

One can write the above coordinate transformation as

$$y = Y \sin \omega, \tag{3.7.3}$$
$$x = X + Y \cos \omega, \tag{3.7.4}$$
$$\Rightarrow \mathbf{x} := \hat{\mathbf{i}} x + \hat{\mathbf{j}} y = \hat{\mathbf{i}} (X + Y \cos \omega) + \hat{\mathbf{j}} Y \sin \omega. \tag{3.7.5}$$

Thus we get the (covariant) basis for the contravariant components as

$$\mathbf{e}_X = \frac{\partial \mathbf{x}}{\partial X} = \hat{\mathbf{i}} \tag{3.7.6}$$

$$\mathbf{e}_Y = \frac{\partial \mathbf{x}}{\partial Y} = \hat{\mathbf{i}} \cos \omega + \hat{\mathbf{j}} \sin \omega, \tag{3.7.7}$$

The (contravariant) basis for the covariant components can be expressed as

$$\mathbf{e}^X = \nabla X = \hat{\mathbf{i}} - \hat{\mathbf{j}} \cot \omega \tag{3.7.8}$$
$$\mathbf{e}^Y = \nabla Y = \hat{\mathbf{j}}/\sin \omega . \tag{3.7.9}$$

Following the above expressions, the direction of the basis vectors have been marked in Fig. 3.2. Although the directions have been marked, the units along different axes are different, as reflected in the amplitude of the basis vectors. For example, unit of length along $\mathbf{e}^Y$ is $|\mathbf{e}^Y| = 1/\sin \omega$, which is reflecting the fact that unit change in Y corresponds to $|\mathbf{e}^Y|$ change in the Cartesian coordinates. Similarly, unit change along the covariant X axis (which we will call x_X), is equivalent to $|\mathbf{e}_X| = 1/\sin \omega$ change in the Cartesian coordinates. It is easy to see that $\mathbf{e}^X \cdot \mathbf{e}_X = \mathbf{e}^Y \cdot \mathbf{e}_Y = 1$ and $\mathbf{e}^X \cdot \mathbf{e}_Y = \mathbf{e}^Y \cdot \mathbf{e}_X = 0$.

For an arbitrary point P, with Cartesian position vector $\mathbf{x} := \hat{\mathbf{i}}x + \hat{\mathbf{j}}y$, the contravariant and covariant components with respect to the tilted coordinate system are given by

$$x^X = \mathbf{x} \cdot \mathbf{e}^X = x - y \cot \omega = X , \tag{3.7.10}$$

$$x^Y = \mathbf{x} \cdot \mathbf{e}^Y = y/\sin \omega = Y , \tag{3.7.11}$$

$$x_X = \mathbf{x} \cdot \mathbf{e}_X = x = X + Y \cos \omega , \tag{3.7.12}$$

$$x_Y = \mathbf{x} \cdot \mathbf{e}_Y = x \cos \omega + y \sin \omega = X \cos \omega + Y . \tag{3.7.13}$$

These are in the units of the corresponding basis vectors, $\mathbf{e}^X$, $\mathbf{e}^Y$, $\mathbf{e}_X$ and $\mathbf{e}_Y$ respectively (not the ones which were used for scalar product). The norm of the vector,

$$x^X x_X + x^Y x_Y = x^2 + y^2 = |\mathbf{x}|^2 , \tag{3.7.14}$$

is indeed consistent with the corresponding Cartesian value.

Now let us represent the contravariant and covariant components in Fig. 3.2. One can do this is two ways. First, the easiest, express $\mathbf{e}_X x^X$, $\mathbf{e}_Y x^Y$, $\mathbf{e}^X x_X$ and $\mathbf{e}^Y x_Y$ in terms of x, y, $\hat{\mathbf{i}}$ and $\hat{\mathbf{j}}$ and plot on the figure, which will show the respective vectors can be represented by OA, OB, OC and OD. Note that, the vectors appear to have different lengths from the actual coordinates values due to the non-unit length of the basis vectors.

One can also mark the components along the $Y = 0$ and $X = 0$ "axes". The contravariant components would by construction be at the points A and B respectively. While the covariant components would be at the foot of the perpendicular drawn from P on those axes, namely, E and F respectively. Thus, if the components are the distances measured along the axes from the origin, the contravariant components are the distances to the points on the axes intercepted by lines drawn from P parallel to those axes, while the covariant components are the distances to the foot of the perpendiculars drawn from P on those axes.

Finally, we look at the metric in this oblique coordinate system. Clearly, from the coordinate transformation formulae, one could write,

$$ds^2 = dx^2 + dy^2 = dX^2 + 2\cos\omega \, dX \, dY + dY^2 . \tag{3.7.15}$$

Thus the metric tensor,

$$g_{ij} = \begin{bmatrix} 1 & \cos\omega \\ \cos\omega & 1 \end{bmatrix}, \tag{3.7.16}$$

is not diagonal. It is trivial to check that the metric can also be derived using the formalism developed earlier, that is, $g_{ij} = \mathbf{e}_i \cdot \mathbf{e}_j$, by substituting the expressions for $\mathbf{e}_X$ and $\mathbf{e}_Y$. Similarly, using the expressions for $\mathbf{e}^X$ and $\mathbf{e}^Y$ and $g^{ij} = \mathbf{e}^i \cdot \mathbf{e}^j$ one could show,

$$g^{ij} = \frac{1}{\sin^2\omega} \begin{bmatrix} 1 & -\cos\omega \\ -\cos\omega & 1 \end{bmatrix}, \tag{3.7.17}$$

which is the inverse of g_{ij}. [2]

3.8   Transition to Curved Spaces

We can readily generalise the foregoing discussion to tensors on curved spaces or Riemannian manifolds. A Riemannian space, roughly speaking, can be thought of as a n dimensional generalisation of a two dimensional surface on which a metric has been defined. Any Riemannian space can be embedded in a sufficiently high dimensional Euclidean space—the maximum dimension required is $n(n + 1)$. The same result also holds for a pseudo-Riemannian space of given signature which can be embedded in a flat space with appropriate signature (Eisenhart (1926)). For example, a 3-sphere can be embedded in the 4-dimensional Euclidean space—one does not need 6 dimensions because it happens to be a highly symmetric space. Consider a 4 dimensional Euclidean space with Cartesian coordinates x^i, $i = 1, 2, 3, 4$, then the equation $(x^1)^2 + (x^2)^2 + (x^3)^2 + (x^4)^2 = 1$ describes a 3 dimensional unit sphere embedded in 4 dimensions. Another example is the 4 dimensional Schwarzschild spacetime which can be embedded in a 6 dimensional flat space (Eisenhart (1926)). If one thinks of a curved space as a subspace of a higher dimensional flat space, then

---

[2] For the advanced reader: Contravariant vectors (tangent vectors) and covariant vectors (1-forms) belong to different vector spaces, namely, the tangent space and the cotangent space respectively. However, the metric tensor defines an isomorphism between them by $V_i = g_{ij} V^j$ and its inverse $V^i = g^{ij} V_j$. This has been demonstrated pictorially in the example.

it inherits the metric - the induced metric—from that flat space. See Exercise 3.10 for a 2 dimensional space (surface) embedded in 3 dimensional Euclidean space. Since we did not put any restriction on dimension of the spaces, all the foregoing discussion on vectors and tensors is valid for n dimensional Riemannian manifolds.

However, embedding the curved space in a higher dimensional Euclidean space is not necessary—we can have a Riemannian space in its own right. We only need to prescribe a set of n coordinates x^i or a coordinate system in general on a subset of the space with sufficiently smoothness properties. This is called a chart. We generally require several charts to cover the space. In fact a curved space cannot be covered by a single chart. Now suppose we have another chart x'^j, then on the intersection of the two charts, we will have x'^is as functions of x^j and vice-versa. In the intersection region of the two charts, the vectors and tensors follow exactly the same transformation laws as discussed earlier. Our analysis therefore is local and restricted to a chart or to an intersection region of charts. In this introductory book, we will restrict ourselves to local analysis. What about the metric? The metric must be prescribed. Since we do not have any Euclidean space in the background, the metric has to be given. It may be positive definite or indefinite as may be required from physical considerations. In relativity, both special and general, we have a 4 dimensional manifold on which the metric is indefinite.

Given these considerations, the foregoing discussion on vectors and tensors applies equally well to general n dimensional Riemannian spaces.

Exercises

1. The position vector in spherical polar coordinate system is given by:

$$\mathbf{x} = r\ \sin\theta\cos\phi\ \hat{\mathbf{i}} + r\ \sin\theta\sin\phi\ \hat{\mathbf{j}} + r\ \cos\theta\ \hat{\mathbf{k}},$$

   where $\hat{\mathbf{i}}, \hat{\mathbf{j}}, \hat{\mathbf{k}}$ are unit vectors along the Cartesian coordinate axes x, y, z respectively. Using Eqs. (3.2.2) and (3.2.3) compute the basis vectors $\mathbf{e}_r, \mathbf{e}_\theta, \mathbf{e}_\phi$ and their reciprocals $\mathbf{e}^r, \mathbf{e}^\theta, \mathbf{e}^\phi$. Verify that they are not all unit vectors, but they satisfy the reciprocity condition Eq. (3.2.5).

2. Let A^i be a contravariant vector and B_i a covariant vector then using the transformation law, show $A^i B_i$ is a scalar.

3. Consider a 2-dimensional surface parametrised by the coordinates (u, v). Let (u', v') be two functions of (u, v), that is, $u'(u, v)$ and $v'(u, v)$. Show that u', v' are not a valid coordinates in a neighbourhood of the point P, if the Jacobian of the transformation vanishes at P. Show that the tangents to the u' and v' curves at P are parallel, so that the curves fail to span a 2-dimensional space in the neighbourhood of P.

4. Let A_i be a covariant vector, then show using the transformation law that $F_{ik} = A_{i,k} - A_{k,i}$ is a covariant tensor of second rank.

5. Quotient law:

(a) Consider the inner product $C_i = A_{ij}B^j$, where C_i is a covariant vector and B^j is an *arbitrary* contravariant vector. Then show that A_{ij} is a covariant tensor of the second rank.

(b) Let $\psi = A_{ij}B^{ij}$ be a scalar, where B^{ij} is a symmetric tensor but otherwise arbitrary. Then show that the symmetric part of A_{ij} is a tensor, namely, $A_{ij} + A_{ji}$.

6. *Physically understanding a second rank tensor:* Consider an anisotropic dielectric material described by the susceptibility χ. Electric field **E** is applied to the dielectric material which polarises it. The polarisation is described by the polarisation vector **P**. For an isotropic material we have the relation $\mathbf{P} = \chi \mathbf{E}$, where χ is a scalar. But if the dielectric material is anisotropic, **P** in general, is not in the same direction as **E**. See Fig. 3.3 below:

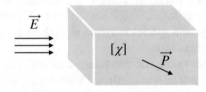

Fig. 3.3 Electric field **E** applied to an anisotropic material produces a polarisation vector **P**, in general, in a different direction. So we must write, $P^i = \chi^i_k E^k$, where the susceptibility is described by a second rank tensor χ^i_k

We must then describe the situation by the equation $P^i = \chi^i_k E^k$, where we have chosen a coordinate system in which we have expressed the quantities **P** and **E** in terms of their components. Then χ^i_k is a second rank tensor, because we may apply arbitrary electric fields **E** to the material and we can invoke the quotient law to prove that χ^i_k is a mixed tensor of second rank. Since χ^i_k is a property of the dielectric material it can be thought of as a single entity $\overset{\leftrightarrow}{\chi}$, just as we regard **P** or **E** as single entities by themselves. In a primed coordinates, show that χ^i_k transforms as a mixed tensor of the second rank.

7. Show that if A_{ij} is symmetric in one coordinate system it is symmetric in all coordinate systems.

8. Consider the metric in standard spherical polar coordinates r,θ,ϕ:

$$ds^2 = dr^2 + r^2(d\theta^2 + \sin^2\theta d\phi^2)$$

Using the standard coordinate transformation equations for tensors, show that the metric tensor transforms to an identity matrix in Cartesian coordinates with the same origin. [Hint: here transforming g^{ik} may be easier than transforming g_{ik}.] Write down the corresponding Jacobian matrix for the transformations.

9. A hyperboloid of revolution is given by the parametric equations:

$$x = a \cosh u \, \cos\phi, \ y = a \cosh u \, \sin\phi, \ z = a \sinh u.$$

Obtain the metric on the hyperboloid in the form:

$$ds^2 = a^2 \cosh 2u \, du^2 + a^2 \cosh^2 u \, d\phi^2.$$

10. A 2-surface is defined by two parametrisations, $x = x(u, v)$, $y = y(u, v)$, $z = z(u, v)$ in Euclidean 3-space with Cartesian coordinates (x, y, z).

(a) Write the metric on the surface in the form:

$$ds^2 = dx^2 + dy^2 + dz^2 \equiv E \, du^2 + 2 \, F \, du \, dv + G \, dv^2,$$

where E, F, G are functions of u, v. This is called the *I*st *Fundamental Form*. It is the *induced* metric the surface inherits from the Euclidean metric.

(b) Let u' and v' be another parametrisation of the same surface so that u', v' are functions of u, v and vice-versa. Then:

$$ds^2 = E' \, du'^2 + 2 \, F' \, du' \, dv' + G' \, dv'^2$$

where E', F', G' are functions of u' and v'. From the arbitrariness of the choice of (du, dv), show that:

$$E' = E \left(\frac{\partial u}{\partial u'}\right)^2 + 2 \, F \left(\frac{\partial u}{\partial u'} \frac{\partial v}{\partial u'}\right) + G \left(\frac{\partial v}{\partial u'}\right)^2.$$

Obtain analogous expressions for F', G'. Recognise that these relations are nothing but the tensor transformation law for the metric tensor. The quotient law has been applied by using the arbitrariness of $dx^i = (du, dv)$ and further using the property that ds^2 is a scalar.

Chapter 4
The Geometry of Curved Spaces and Tensor Calculus

4.1 Introduction

Comparison of two vectors at two different points of a manifold is non-trivial in curvilinear coordinates. In flat space, we are used to comparing two vectors at two different locations. For this purpose we naturally use the Cartesian coordinate system. For example, we define a constant electric field $\mathbf{E}$ as having the same components (E_x, E_y, E_z) at every location (x, y, z). In fact, we define the constancy of all the Cartesian components as a necessary and sufficient condition for the constancy of a vector field. However, if we go over to curvilinear coordinates our usual notion of constancy of components becomes ambiguous. For example, in polar coordinates on a plane, if a vector field $\mathbf{E}$ has a constant radial component $E_r = E$ and the azimuthal component $E_\theta = 0$ at every (r, θ), the vector field $\mathbf{E}$ is not a constant vector field according to our usual understanding, as it points in different directions for different values of θ—the field is radial. If we had chosen some other coordinate system (different from Cartesian and polar), we would have arrived at a different result. Conversely, a constant vector field $\mathbf{E}$, along the "x-axis" ($\theta = 0$ line) will have different components at different positions in space $E_r = E \sin \theta$ and $E_\theta = E \cos \theta$. Figure 4.1 provides a pictorial explanation. Thus to compare two vectors at two different locations in curvilinear coordinates, one needs the notion of "parallel transport"—we need to *define* a vector at $Q(x^k + dx^k)$ say, parallel to a given vector at $P(x^k)$—it is done most conveniently for a small displacement dx^k. Parallel transport defined by the constancy of Cartesian components of a vector field leads to the notion of *Riemannian* parallel transport. Other types of parallel transports can be defined also for various other applications. In this book we will confine ourselves to Riemannian parallel transport which as we will see is defined via the metric.

The original version of this chapter was revised: Belated corrections in equations have been updated. The correction to this chapter is available at https://doi.org/10.1007/978-3-030-92335-8_10

© The Author(s), under exclusive license to Springer Nature Switzerland AG 2022, corrected publication 2022
S. Dhurandhar and S. Mitra, *General Relativity and Gravitational Waves*, UNITEXT for Physics, https://doi.org/10.1007/978-3-030-92335-8_4

Fig. 4.1 This illustration
shows how a constant
electric field **E** can have
different radial components
at different positions in a
curvilinear coordinates,
two-dimensional polar
coordinates in this case. The
radial component of **E** at
position A(r, θ) is
$E_r = E \cos \phi$, but at
B$(r, \theta + \delta\theta)$ is
$E_r = E \cos(\phi - \delta\theta)$

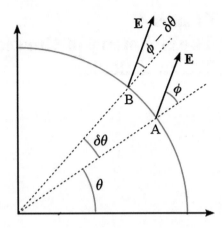

The primary reason for defining parallel transport is to define derivatives of tensors in a consistent way: that is we demand that the derivative of a tensor should also be a tensor. This demand arises primarily from physics in which we require coordinate independent quantities and tensors are coordinate independent quantities. Parallel transport achieves this by making it possible to subtract two tensors at the *same* point, because then the difference is also a tensor—the operation of subtraction is in the same vector space. For example for a vector, there is no meaning to subtracting vectors at two different points as they belong to different tangent spaces—the operation is undefined! It cannot be done.

The discussion which follows, will make these points clear.

4.2 Parallel Transport and the Covariant Derivative

To gain insight into the procedure let us first take the simple case of a vector field with contravariant components, $A^i(x^k) \equiv (A^r(r, \theta), A^\theta(r, \theta))$ defined on a plane in polar coordinates. Our goal is to parallely transport this vector to the point $(r + dr, \theta + d\theta)$.

Note that we are in a plane and on which we have the luxury of a Cartesian system (x, y) which covers the plane. We can use this coordinate system to achieve our end. We must transform the polar components of the vector (A^r, A^θ) to Cartesian components (A^x, A^y). Then use the constancy of Cartesian components (A^x, A^y) to transport it to the neighbouring point given by the coordinates $(r + dr, \theta + d\theta)$. We have the following transformation equations between the coordinates:

$$x = r \cos \theta, \quad y = r \sin \theta; \quad r = \sqrt{x^2 + y^2}, \quad \theta = \tan^{-1} \frac{y}{x}. \tag{4.2.1}$$

Using the transformation law for contravariant components, we have:

$$A^r = \frac{\partial r}{\partial x} A^x + \frac{\partial r}{\partial y} A^y = \cos\theta \, A^x + \sin\theta \, A^y \tag{4.2.2}$$

$$A^\theta = \frac{\partial \theta}{\partial x} A^x + \frac{\partial \theta}{\partial y} A^y = \frac{1}{r} \left(-\sin\theta \, A^x + \cos\theta \, A^y \right). \tag{4.2.3}$$

Now the constancy of Cartesian components implies that under parallel transport $\delta A^x = \delta A^y = 0$—there is no change in the Cartesian components. Using this fact and the above equations we readily obtain:

$$\delta A^r = (-A^x \sin\theta d\theta + \cos\theta \, A^y d\theta) = r A^\theta d\theta, \tag{4.2.4}$$

$$\delta A^\theta = -\frac{1}{r} A^\theta dr - \frac{1}{r} A^r d\theta. \tag{4.2.5}$$

Thus the parallely transported vector at $(r + dr, \theta + d\theta)$ is $(A^r + \delta A^r, A^\theta + \delta A^\theta)$ where the changes in the components are given by the above equations and is parallel to the former vector. It preserves the constancy of Cartesian components, i.e. $A^x(r, \theta) = A^x(r + dr, \theta + d\theta)$ and $A^y(r, \theta) = A^y(r + dr, \theta + d\theta)$.

We could parallely transport the vector as above because we could fall back on the Cartesian coordinate system which was available in the plane because it was a flat space. But what about in curved spaces such as a sphere? Here we do not have a Cartesian coordinate system to refer to. The usual spherical coordinate system (θ, ϕ) is not Cartesian, because the metric components are not constants. So then how can we solve the problem? The answer is that we can still parallely transport a vector *but only locally*, that is for differential displacements. Note that in the example above we could have just as easily parallely transported the vector for finite displacements. We purposely chose only differential displacements in that example, because we foresaw that we would need to port a similar procedure on curved spaces. Observe that the sphere is locally flat and it is possible to choose a local Cartesian coordinate system in a small patch. But what is a local Cartesian system? Note that a Cartesian system is one for which the metric components are constants. Accordingly, we define a local Cartesian system to be one in which the metric components g_{ij} are locally constants. More specifically, in order to have local Cartesian system around a point P say, we demand that $(g_{ij,k})_P = 0$, where the comma denotes partial derivative with respect to the coordinate x^k. We assert that we can always do this in a Riemannian manifold. This is argued later in Sect. 4.4. We may sometimes refer to the local Cartesian system also as locally flat.

Below we give the general procedure which carries out parallel transport in a Riemannian manifold by taking a vector A^i at a point $P(x^k)$ in the manifold and parallely transporting this vector to a nearby point $Q(x^k + dx^k)$.

- Choose a locally Cartesian system around P such that $(g_{ij,k})_P = 0$.
- Transform the components of the vector A^i from the general coordinate system x^i to a local Cartesian system say x'^i so that we have the quantities A'^i.

- Use the constancy of Cartesian components A'^i in the local Cartesian system to parallely transport the vector, that is, set $\delta A'^i = 0$.
- Transform back the Cartesian components of A'^i at Q to the original coordinate system x^i obtaining the components of the parallely transported vector $A^i + \delta A^i$ at Q and express all quantities as functions of x^i. Note that in general $\delta A^i \neq 0$.

Starting from the relation:

$$A^i = \frac{\partial x^i}{\partial x'^j} A'^j , \tag{4.2.6}$$

following the above procedure yields the result:

$$\delta A^i = \frac{\partial^2 x^i}{\partial x'^j \partial x'^m} \frac{\partial x'^j}{\partial x^n} \frac{\partial x'^m}{\partial x^k} A^n dx^k \tag{4.2.7}$$

$$\equiv -\Gamma^i_{nk} A^n dx^k . \tag{4.2.8}$$

The Γ symbols are called Christoffel symbols. Interchanging the indices j and n, they are given by:

$$\Gamma^i_{jk} = -\frac{\partial^2 x^i}{\partial x'^m \partial x'^m} \frac{\partial x'^m}{\partial x^j} \frac{\partial x'^m}{\partial x^k} . \tag{4.2.9}$$

The minus sign is a convention. Immediately we conclude that they must be symmetric in the lower indices.

$$\Gamma^i_{jk} = \Gamma^i_{kj} . \tag{4.2.10}$$

The above procedure is known as *Riemannian* parallel transport and it crucially depends on the metric—we made use of the metric to get local Cartesian coordinates. The Christoffel symbols depend on the metric—they are essentially first derivatives of the metric. In Sect. 4.4 we will obtain an expression for them in terms of the metric. Note that from the expression for δA^i, it is linear both in the vector A^i and the displacement dx^k.

A question that may occur about the above procedure is that, is it "well defined"? This is because there is huge freedom in choosing local Cartesian coordinates at P, and it is possible to choose different local Cartesian systems at P. The question is: does this freedom lead to ambiguity in parallelly transporting vectors, that is, does choosing different local Cartesian systems lead to different parallelly transported vectors? We argue below that this is indeed not the case and the operation of parallel transport defined in this way is well defined. Let x'^i be a local Cartesian system and consider another local Cartrsian system x'''^i, then they are connected by the transformations of the form $x'''^i = \Lambda^i_k x'^k$, where Λ^i_k in general, is a $n \times n$ invertible *constant* matrix. We write the inverse relations as $x'^i = \Xi^i_k x'''^k$, where the matrix Ξ is the inverse of the matrix Λ and is also a constant matrix. One can easily verify that if x'^i is a local Cartesian coordinate system at P, then so is x'''^i. To show this, we just compute:

$$\frac{\partial g''_{ij}}{\partial x'''^k} = \Xi^m_i \, \Xi^n_j \, \frac{\partial g'_{mn}}{\partial x'''^k} = \Xi^m_i \, \Xi^n_j \, \frac{\partial x'^l}{\partial x'''^k} \, g'_{mn,l} = \Xi^m_i \, \Xi^n_j \, \Xi^l_k \, g'_{mn,l} \equiv 0, \tag{4.2.11}$$

where all the derivatives are evaluated at P. This proves that x''^i is also a local Cartesian system. Now consider a contravariant vector A'^i at P in the coordinate system x'^i. Parallelly transporting A'^i to a nearby point Q is achieved by requiring $\delta A'^i = 0$. In the double primed coordinate system we have $\delta A''^i = \Lambda_k^i \, \delta A'^k = 0$, since Λ_k^i is a constant matrix. This shows that the property of constancy of components is preserved also in x''^i, establishing our assertion. Thus given a vector at P, we get the "same" parallelly transported vector at Q irrespective of the local Cartesian system we use, although the components may differ for different local Cartesian coordinate systems.

What about the Christoffel symbols? We see that they remain the same under change of local Cartesian coordinates at P. If we use x''^i to compute the $\Gamma^i{}_{jk}$, then in Eq. (4.2.9), the x'^i is just replaced by x''^i. We can straight forwardly compute the partial derivatives with respect to both sets of coordinates and obtain the relations between them as follows:

$$\frac{\partial x''^n}{\partial x^j} = \frac{\partial x''^n}{\partial x'^p} \frac{\partial x'^p}{\partial x^j} = \Lambda^n_p \frac{\partial x'^p}{\partial x^j}$$

$$\frac{\partial^2 x^i}{\partial x''^n \partial x''^m} = \Xi^p_n \, \Xi^q_m \frac{\partial^2 x^i}{\partial x'^p \partial x'^q} \, . \tag{4.2.12}$$

Since Λ and Ξ are inverses of each other, they cancel out each other, resulting in the same expression for $\Gamma^i{}_{jk}$ as in Eq. (4.2.9). This shows that we may choose any local Cartesian coordinate system to carry out the parallel transport operation and still obtain identical results. The Christoffel symbols $\Gamma^i{}_{jk}$ solely depend on the general coordinate system x^i and not on the choice of the local Cartesian coordinate system at P.

We now go over to define a derivative of a vector or in general tensors of arbitrary rank. In order to define the derivative of a vector we require a vector field say $A^i(x^k)$, where we assume that the components A^i are differentiable functions of x^k as many times as required. Since A^i is a vector field it is defined in the neighbourhood of x^k and we can take the differentials of each of its components:

$$\mathrm{d}A^i = \frac{\partial A^i}{\partial x^j} \, \mathrm{d}x^j \, , \tag{4.2.13}$$

where we have made use of the chain rule. But now we will show that dA^i does not transform as a vector and $\partial A^i / \partial x^j$ does not transform as a tensor. From the index structure one would have expected it to transform as a mixed tensor of the second rank, but it does not. We see this as follows: Let us take the first order differential of transformation equation of A^i:

$$A'^i = A^j \frac{\partial x'^i}{\partial x^j} \tag{4.2.14}$$

$$\Rightarrow \quad \mathrm{d}A'^i = \mathrm{d}A^j \frac{\partial x'^i}{\partial x^j} + A^j \frac{\partial^2 x'^i}{\partial x^k \, \partial x^j} \, \mathrm{d}x^k \, . \tag{4.2.15}$$

So $\mathrm{d}A^i$ transforms as a vector only if the second term on the right is zero; which happens for linear transformations. Also note that the second term contains a second partial derivative which in fact resembles the expression for the Christoffel symbols defined above. This is because $\mathrm{d}A^i$ is indeed not the difference of two vectors at the same point. The fix is obvious. If we transport A^i from x^k to $x^k + \mathrm{d}x^k$ and take the difference between $A^i(x^k + \mathrm{d}x^k)$ and the parallel transported vector, the difference quantity would transform as a vector. So we parallely transport the vector $A^i(x^k)$ to $x^k + \mathrm{d}x^k$ by adding δA^i to $A^i(x^k)$ to obtain $A^i + \delta A^i$, where δA^i is given by,

$$\delta A^i = -\Gamma^i{}_{jk} A^j \, \mathrm{d}x^k . \tag{4.2.16}$$

We may now subtract $A^i + \delta A^i$ from $A^i + dA^i$ and obtain a differential say DA^i which does transform as a vector (the proof will follow), because we are subtracting two vectors at the same point. We therefore have:

$$\mathrm{D}A^i = \mathrm{d}A^i - \delta A^i = \left(\frac{\partial A^i}{\partial x^k} + \Gamma^i{}_{jk} A^j \right) \mathrm{d}x^k . \tag{4.2.17}$$

We call the quantity in the parenthesis, the *covariant derivative* of A^i,

$$A^i_{;k} := A^i_{,k} + \Gamma^i{}_{jk} A^j . \tag{4.2.18}$$

Here we have introduced the shorthand semi-colon ; k to express covariant differentiation with respect to x^k and the ordinary comma , k for partial differentiation. Thus $A^i_{,k} := \partial A^i / \partial x^k$. We may occasionally use "D_k" in place of "; k" and "∂_k" in place of ", k". Since DA^i is a vector (as we will show) and dx^k is an arbitrary vector, from the quotient law, $A^i_{;k}$ must be a mixed tensor of rank 2.

Next, we see how we can covariantly differentiate covariant vectors. Let us now take two arbitrary vectors A^i and B_i, then if we parallel transport both the vectors from $P(x^k)$ to $Q(x^k + dx^k)$, then invoking the local Cartesian coordinates x'^i we see from the constancy of the Cartesian components, that we must have $\delta(A'^i B_i') = 0$. But since the scalar product is invariant under coordinate transformations, we must have $\delta(A^i B_i) = 0$ in a general coordinate system. Thus it follows that,

$$\delta(A^i B_i) = 0 \quad \Rightarrow \quad -A^i \delta B_i = B_i \delta A^i = -B_i \Gamma^i{}_{jk} A^j \, \mathrm{d}x^k. \tag{4.2.19}$$

Then from the arbitrariness of the A^i, one obtains,

$$\delta B_i = \Gamma^j{}_{ik} B_j \, \mathrm{d}x^k . \tag{4.2.20}$$

One may then define the covariant differential and covariant derivative for covariant vectors by the relations:

$$DB_i = dB_i - \delta B_i = \left(\frac{\partial B_i}{\partial x^k} - \Gamma^j_{ik} B_j \right) dx^k, \tag{4.2.21}$$

$$B_{i;k} := B_{i,k} - \Gamma^j_{ik} B_j. \tag{4.2.22}$$

A similar approach for obtaining the covariant derivatives of contravariant and covariant vectors has been followed in Narlikar [2010].

These formulae can be extended to tensors of higher ranks. For example, to obtain the covariant derivative of a mixed second rank tensor T^i_k consider the parallel transport of the quantity $T^i_k A_i B^k$, where A_i and B^k are arbitrary vectors. Since $T^i_k A_i B^k$ is a scalar, $\delta(T^i_k A_i B^k) = 0$. Since we know the expressions for δA_i and δB^k, we can obtain δT^i_k and hence the covariant derivative of the tensor field T^i_k. In this way we obtain:

$$T^i_{j;k} = T^i_{j,k} + \Gamma^i_{mk} T^m_j - \Gamma^m_{jk} T^i_m \tag{4.2.23}$$

$$S_{ij;k} = S_{ij,k} - \Gamma^m_{ik} S_{mj} - \Gamma^m_{jk} S_{im}, \tag{4.2.24}$$

$$R^{ij}_{;k} = R^{ij}_{,k} + \Gamma^i_{mk} R^{mj} + \Gamma^j_{mk} R^{im}. \tag{4.2.25}$$

We may also obtain the product rule for covariant differentiation. Consider, for example, two covariant vectors A_i and B_j, then what is $(A_i B_j)_{;k}$? Let $T_{ij} = A_i B_j$ which is a second rank covariant tensor. Take the covariant derivative of T_{ij} as shown above and substitute for A_i and B_j. We then obtain the result:

$$(A_i B_j)_{;k} = A_{i;k} B_j + A_i B_{j;k}. \tag{4.2.26}$$

One can also raise and lower covariant differentiation indices,

$$A_i^{;j} = g^{jk} A_{i;k}, \tag{4.2.27}$$

and similarly,

$$A_{i;j} = g_{jk} A_i^{;k}. \tag{4.2.28}$$

Next we see how Γ^i_{jk} transform from a coordinate system x^i to another one $x'^{i'}$. Let us consider a local Cartesian coordinate system $x''^{i''}$. Then from Eq. (4.2.9) one can write

$$\Gamma^i_{kl} = -\frac{\partial^2 x^i}{\partial x''^{k''} \partial x''^{l''}} \frac{\partial x''^{k''}}{\partial x^k} \frac{\partial x''^{l''}}{\partial x^l}, \tag{4.2.29}$$

$$\Gamma'^{i'}_{k'l'} = -\frac{\partial^2 x'^{i'}}{\partial x''^{k''} \partial x''^{l''}} \frac{\partial x''^{k''}}{\partial x'^{k'}} \frac{\partial x''^{l''}}{\partial x'^{l'}}. \tag{4.2.30}$$

The task is to relate the equations only through functions of the primed and the unprimed coordinates. One can transform the second equation as

$$\frac{\partial x^i}{\partial x''^{i'}} \frac{\partial x'^{k'}}{\partial x^k} \frac{\partial x'^{l'}}{\partial x^l} \Gamma''^{i'}_{k'l'} = -\frac{\partial x^i}{\partial x''^{i'}} \frac{\partial^2 x''^{i'}}{\partial x''^{k''} \partial x''^{l''}} \frac{\partial x'^{k'}}{\partial x^k} \frac{\partial x''^{k''}}{\partial x'^{k'}} \frac{\partial x'^{l'}}{\partial x^l} \frac{\partial x''^{l''}}{\partial x'^{l'}}$$

$$= -\frac{\partial x^i}{\partial x''^{i'}} \frac{\partial}{\partial x''^{k''}} \left(\frac{\partial x^m}{\partial x''^{l''}} \frac{\partial x''^{i'}}{\partial x^m} \right) \frac{\partial x''^{k''}}{\partial x^k} \frac{\partial x''^{l''}}{\partial x^l}$$

$$= -\delta^i_m \left(\frac{\partial^2 x^m}{\partial x''^{k''} \partial x''^{l''}} \right) \frac{\partial x''^{k''}}{\partial x^k} \frac{\partial x''^{l''}}{\partial x^l} - \frac{\partial x^i}{\partial x''^{i'}} \frac{\partial}{\partial x^k} \left(\frac{\partial x''^{i'}}{\partial x^m} \right) \delta^m_l .$$

Thus one can write

$$\Gamma^i_{kl} = \frac{\partial x^i}{\partial x''^{i'}} \left(\frac{\partial x'^{k'}}{\partial x^k} \frac{\partial x'^{l'}}{\partial x^l} \Gamma''^{i'}_{k'l'} + \frac{\partial^2 x''^{i'}}{\partial x^k \partial x^l} \right) . \qquad (4.2.31)$$

Clearly, Γ^i_{kl} does not transform as a tensor because of the second term on the right. Note however that $\Gamma^i_{kl} - \Gamma^i_{lk}$ does transform like a tensor, as the second term on the right cancels out. Which implies that if the Christoffel symbols are symmetric about the lower indices in one coordinate system, they necessarily stay symmetric in every other coordinate system. Which is of course inbuilt in the way we have prescribed parallel transport.

We will now explicitly obtain an expression for the Christoffel symbols in terms of the metric. We start with the equation $\delta(A^i B_i) = 0$ where the δ represents the operation of parallel transport. We parallel transport the vectors A^i and B^i from x^k to $x^k + dx^k$ and equate their scalar products $A_i B^i = g_{ij} A^i B^j$ at these two points. We write this equation as:

$$\delta(g_{ij} A^i B^j) = g_{ij,k} A^i B^j dx^k + g_{ij} \delta A^i B^j + g_{ij} A^i \delta B^j = 0. \qquad (4.2.32)$$

The first term is obtained by δ operating on a scalar, while for the latter two terms we use the parallel transport law in terms of Christoffel symbols. We write δA^i and δB^j in terms of the Christoffel symbols using Eq. (4.2.16) and then after changing several dummy indices obtain:

$$\left(g_{ij,k} - g_{mj} \Gamma^m_{ik} - g_{im} \Gamma^m_{jk} \right) A^i B^j dx^k = 0. \qquad (4.2.33)$$

Since A^i, B^j and dx^k are arbirtary we obtain:

$$g_{ij;k} \equiv g_{ij,k} - \Gamma^m_{ik} g_{mj} - \Gamma^m_{jk} g_{im} = 0. \qquad (4.2.34)$$

Cyclically permuting the indices i, j and k, we obtain two more equations:

$$g_{jk,i} - \Gamma_{kji} - \Gamma_{jki} = 0, \qquad (4.2.35)$$

$$g_{ki,j} - \Gamma_{ikj} - \Gamma_{kij} = 0, \qquad (4.2.36)$$

where we have defined the symbol Γ_{ijk} which is also a Christoffel symbol but of the *first kind* (Eisenhart [1926]). Adding the first and third equation and subtracting the second we obtain,

$$\Gamma_{ijk} = \frac{1}{2}(g_{ij,k} + g_{ik,j} - g_{jk,i}). \tag{4.2.37}$$

Raising the first index we obtain,

$$\Gamma^i{}_{jk} = \frac{1}{2}g^{im}(g_{mj,k} + g_{mk,j} - g_{jk,m}), \tag{4.2.38}$$

which is the Christoffel symbol of the *second kind* (Eisenhart [1926]). The two Christoffel symbols are related to each other by the equations:

$$\Gamma_{ijk} := g_{im}\Gamma^m{}_{jk}, \qquad \Gamma^i{}_{jk} = g^{im}\Gamma_{mjk}. \tag{4.2.39}$$

Often Eq. (4.2.34) in differential geometric language is described by the sentence: *the affine connection is compatible with the metric.*

We also saw in Eq. (4.2.15) that dA^i also does not transform like a vector. However, DA^i does transform as a vector. We show this below. Substituting Eq. (4.2.17) in Eq. (4.2.15), one gets

$$dA'^i = \frac{\partial x'^i}{\partial x^k}dA^k + \frac{\partial^2 x'^i}{\partial x^k \partial x^l}A^l dx^k$$

$$\Rightarrow DA'^i + \delta A'^i = (DA^k + \delta A^k)\frac{\partial x'^i}{\partial x^k} + \frac{\partial^2 x'^i}{\partial x^k \partial x^l}A^l dx^k.$$

Since,

$$\delta A^m \frac{\partial x'^i}{\partial x^m} = -\Gamma^m_{kl}A^l dx^k \frac{\partial x'^i}{\partial x^m}$$

$$= -\left(\frac{\partial x^m}{\partial x'^n}\frac{\partial x'^p}{\partial x^k}\frac{\partial x'^q}{\partial x^l}\Gamma'^n_{pq} + \frac{\partial^2 x'^n}{\partial x^k \partial x^l}\frac{\partial x^m}{\partial x'^n}\right)A^l dx^k \frac{\partial x'^i}{\partial x^m}$$

$$= -\Gamma'^i_{kl}A^l dx^k - \frac{\partial^2 x'^i}{\partial x^k \partial x^l}A^l dx^k = \delta A'^i - \frac{\partial^2 x'^i}{\partial x^k \partial x^l}A^l dx^k, \tag{4.2.40}$$

Substituting this transformation law for δA^i in the above equation cancels out this quantity completely as well as the second partial derivative term. Then one is left with,

$$DA'^i = \frac{\partial x'^i}{\partial x^m}DA^m, \tag{4.2.41}$$

which proves that DA^i is a vector. We can bring the index down and obtain,

$$(DA)_i = g_{ij}DA^j = D(g_{ij}A^j) = D(A_i). \tag{4.2.42}$$

The second equality follows from $g_{ij;k} = 0$.

Derivation 6 *Find the Christoffel symbols for a two-sphere of radius R.*

The metric on the sphere is:

$$ds^2 = R^2 (d\theta^2 + \sin^2\theta \, d\phi^2),\tag{4.2.43}$$

and the metric tensor g_{ij} in the matrix form is given by:

$$g_{ij} = \begin{pmatrix} R^2 & 0 \\ 0 & R^2 \sin^2\theta \end{pmatrix}, \quad g^{ij} = \begin{pmatrix} 1/R^2 & 0 \\ 0 & 1/(R^2 \sin^2\theta) \end{pmatrix}.\tag{4.2.44}$$

Since the only non-zero derivative of g_{ij} is $g_{\phi\phi,\theta} = 2R^2 \sin\theta \cos\theta$, the non-zero Christoffel symbols follow:

$$\Gamma_{\theta\phi\phi} = -(1/2) \, g_{\phi\phi,\theta} \qquad\qquad = -R^2 \sin\theta \cos\theta,\tag{4.2.45}$$

$$\Gamma_{\phi\theta\phi} = \Gamma_{\phi\phi\theta} = (1/2) \, g_{\phi\phi,\theta} \qquad = R^2 \sin\theta \cos\theta,\tag{4.2.46}$$

$$\Gamma^{\theta}{}_{\phi\phi} = g^{\theta\theta} \, \Gamma_{\theta\phi\phi} \qquad\qquad = -\sin\theta \cos\theta\tag{4.2.47}$$

$$\Gamma^{\phi}{}_{\theta\phi} = \Gamma^{\phi}{}_{\phi\theta} = g^{\phi\phi} \, \Gamma_{\phi\phi\theta} \qquad = \cot\theta.\tag{4.2.48}$$

The following identities are often useful:

$$\Gamma^i{}_{ij} = \frac{\partial}{\partial x^j} \ln\sqrt{|g|},\tag{4.2.49}$$

$$g^{jk} \, \Gamma^i{}_{jk} = -\frac{1}{\sqrt{|g|}} \frac{\partial(\sqrt{|g|}\, g^{ik})}{\partial x^k},\tag{4.2.50}$$

$$A^i{}_{;i} = \frac{1}{\sqrt{|g|}} \frac{\partial(\sqrt{|g|}\, A^i)}{\partial x^i},\tag{4.2.51}$$

$$F^{ij}{}_{;j} = \frac{1}{\sqrt{|g|}} \frac{\partial(\sqrt{|g|}\, F^{ij})}{\partial x^j} \quad \text{for, } F^{ji} = -F^{ij}.\tag{4.2.52}$$

Derivation 7 *Show that: $\Gamma^i{}_{ij} = \frac{\partial}{\partial x^j} \ln\sqrt{|g|}$.*

From Eq. (4.2.38),

$$\Gamma^i{}_{ij} = \frac{1}{2} g^{im} (g_{mj,i} + g_{mi,j} - g_{ji,m}).\tag{4.2.53}$$

Interchanging dummy indices i and m in the last term, using symmetry of the metric tensor and finally using Eq. (3.6.4),

$$\Gamma^i{}_{ij} = \frac{1}{2} g^{im} g_{mi,j} \cdot = \frac{\partial}{\partial x^j} \ln\sqrt{|g|}.\tag{4.2.54}$$

Derivation 8 *Show that:* $g^{jk} \Gamma^i{}_{jk} = -\dfrac{1}{\sqrt{|g|}} \dfrac{\partial(\sqrt{|g|}\, g^{ik})}{\partial x^k}$.

From Eq. (4.2.38),

$$g^{jk}\Gamma^i{}_{jk} = \frac{1}{2} g^{jk} g^{im} (g_{mj,k} + g_{mk,j} - g_{jk,m}) = g^{jk} g^{im} \left(g_{mj,k} - \frac{1}{2} g_{jk,m}\right).$$

Since $(g^{im} g_{mj})_{,k} = 0$, *one gets* $g^{im} g_{mj,k} = -g^{im}_{,k} g_{mj}$. *Then using Eq. (3.6.4)*

$$g^{jk}\Gamma^i{}_{jk} = -g^{jk} g^{im}_{,k} g_{mj} - \frac{1}{2} g^{im} g^{jk} g_{jk,m} = -g^{im}_{,m} - g^{im} \frac{1}{\sqrt{|g|}}(\sqrt{|g|})_{,m}$$

$$= -\frac{1}{\sqrt{|g|}}(\sqrt{|g|}\, g^{im}_{,m} + g^{im}(\sqrt{|g|})_{,m}) = -\frac{1}{\sqrt{|g|}} \frac{\partial(\sqrt{|g|}\, g^{ik})}{\partial x^k}.$$

Derivation 9 *Show that:* $A^i{}_{;i} = \dfrac{1}{\sqrt{|g|}} \dfrac{\partial(\sqrt{|g|}\, A^i)}{\partial x^i}$.

From Eqs. (4.2.18) and (4.2.49) and changing dummy index from j to i,

$$A^i{}_{;i} := A^i{}_{,i} + \Gamma^i{}_{ji} A^j = A^i{}_{,i} + A^i \frac{1}{\sqrt{|g|}} \frac{\partial}{\partial x^i} \sqrt{|g|} = \frac{1}{\sqrt{|g|}} \frac{\partial(\sqrt{|g|}\, A^i)}{\partial x^i}.$$

Derivation 10 *Show that:* $F^{ij}{}_{;j} = \dfrac{1}{\sqrt{|g|}} \dfrac{\partial(\sqrt{|g|}\, F^{ij})}{\partial x^j}$ *for,* $F^{ji} = -F^{ij}$.

From Eq. (4.2.25),

$$F^{ij}{}_{;j} = F^{ij}{}_{,j} + \Gamma^i{}_{mj} F^{mj} + \Gamma^j{}_{mj} F^{im}. \tag{4.2.55}$$

The second term is a contraction of symmetric and anti-symmetric tensors. By interchaging dummy indices and using anti/symmetry condition, the term becomes negative of itself, hence it is zero:

$$\Gamma^i{}_{mj} F^{mj} = \Gamma^i{}_{jm} F^{jm} = (\Gamma^i{}_{mj})(-F^{mj}) = -\Gamma^i{}_{mj} F^{mj}; \quad \Rightarrow 2\,\Gamma^i{}_{mj} F^{mj} = 0. \tag{4.2.56}$$

Then, changing dummy index from m to j,

$$F^{ij}{}_{;j} = F^{ij}{}_{,j} + \Gamma^j{}_{mj} F^{im} = F^{ij}{}_{,j} + F^{im} \frac{1}{\sqrt{|g|}} \frac{\partial}{\partial x^m} \sqrt{|g|} = \frac{1}{\sqrt{|g|}} \frac{\partial(\sqrt{|g|}\, F^{ij})}{\partial x^j}. \tag{4.2.57}$$

## 4.3	Geodesics

In curved spaces, the "straight lines" are called geodesics. Of course, geodesics may not look straight when they are embedded in a flat higher dimensional space. For example, great circles are geodesics on a two-sphere. For a tiny ant moving on the sphere, a geodesic is indeed like a straight line. But when the two sphere is embedded in 3 dimensional Euclidean space, they trace out circles, and that is why we write "great circles are geodesics on a two-sphere".

We now ask the question how would an ant on a general curved surface draw geodesics? Extending our perception of Euclidean geometry, we can think of two ways:

1. **Parallelly transporting the tangent:** The ant takes the first step in a given direction and keeps on moving in the same direction in such a way that the tangent keeps the same direction, that is, the tangent is parallelly transported.
2. **Shortest Distance:** The ant makes large number (tending to infinity) of trips to another point via all possible routes and, calls the route with shortest distance, requiring minimum number of steps, as the geodesic.[1]

When the metric is positive definite as it is for a sphere, the geodesic between two points gives the shortest distance as compared with curves in its neighbourhood. But when the metric is not positive definite as we have in general relativity, the geodesic gives the extremal distance. For example if the two points are timelike separated, the geodesic gives the maximum distance. However, the first of the criterion remains the same even for an indefinite metric.

Let a geodesic be parameterised by a the variable λ. One can consider the λ's to be milestones, which are not unique, in the sense that the locations of the $\lambda = 0, 1, 2, \cdots$ marks depend on the units (km, miles etc.), nevertheless in given units λ uniquely specifies a point on the curve. We shall derive the equations for the geodesic curves in terms of the parameter λ.

### 4.3.1	Parallelly Transporting the Tangent

The components of the tangent vector u^i of a curve $x^i(\lambda)$ is,

$$u^i = \mathrm{d}x^i/\mathrm{d}\lambda. \tag{4.3.1}$$

(since each x^i is a function of λ only!). Parallelly transporting the tangent gives,

---

[1] If the ants were equipped with laser light for two dimensions, one ant will try to send a laser beam towards another ant at a second point by slowly changing the direction. As soon as the second ant receives it, it reflects it back to the first ant by slowly rotating a mirror. The path of the laser can be traced by putting intermediate observers and called the geodesic.

$$u^i_{;k} u^k = 0.$$ (4.3.2)

Since this differential is along a curve parameterised by λ, the above equation can be written as

$$\frac{\partial}{\partial x^k}\left(\frac{dx^i}{d\lambda}\right)\frac{dx^k}{d\lambda} + \Gamma^i_{jk} u^j u^k = \frac{d^2 x^i}{d\lambda^2} + \Gamma^i_{jk}\frac{dx^j}{d\lambda}\frac{dx^k}{d\lambda} = 0.$$ (4.3.3)

This second order differential equation is called the *geodesic equation*. The solutions to this differential equation give the geodesics or geodesic curves.

4.3.2 Extremal Distance

The distance between two points A and B along a curve parameterised by λ is given by

$$\Delta s_{AB} = \int_A^B ds = \int_A^B d\lambda \sqrt{g_{ij}\frac{dx^i}{d\lambda}\frac{dx^j}{d\lambda}},$$ (4.3.4)

where ds is the differential distance along a trajectory. The task is to find the curve $x^i(\lambda)$ that extremises the distance Δs_{AB}. From the calculus of variation, we know that this problem is equivalent of extremising the action,

$$\mathcal{A} := \int_A^B L[x^i(\lambda), \dot{x}^i(\lambda), \lambda]\, d\lambda,$$ (4.3.5)

where the Lagrangian L is given by,

$$L[x^i, \dot{x}^i, \lambda] := \sqrt{g_{ij}\, \dot{x}^i \dot{x}^j},$$ (4.3.6)

is a functional of the trajectory $x^i(\lambda)$ and the "dot" represents differentiation with respect to λ, that is, $\dot{x}^i = dx^i/d\lambda$. The geodesic can then be obtained by solving the Euler–Lagrange equations:

$$\frac{d}{d\lambda}\left(\frac{\partial L}{\partial \dot{x}^i}\right) = \frac{\partial L}{\partial x^i}.$$ (4.3.7)

If we choose the parameter λ to be proportional to the arc-length s—it is called an *affine parameter* (the term affine parameter is explained later in the text)—as they are the natural "milestones". Then clearly along a trajectory $L = $ constant (in space-time, for null curves, $L = 0$), that is, the Lagrangian is a constant of motion.[2] In cases where the Lagrangian L is a constant of motion, one can use any monotonic

---

[2] Of course, there is nothing wrong if the Lagrangian is a constant of motion along a given trajectory. The task is to extremise the action, which is understandably $\int_A^B ds$ in our case. With the calculus

function of L, say $f(L)$ as the Lagrangian, the proof is as follows. If one extremises the action $\int_A^B f(L)\,d\lambda$, the Lagrange's equation becomes

$$\frac{d}{d\lambda}\left(\frac{df}{dL}\frac{\partial L}{\partial \dot{x}^i}\right) = \frac{df}{dL}\frac{\partial L}{\partial x^i}, \tag{4.3.8}$$

Then if L is a constant of motion, so is $f(L)$ and df/dL and hence $dL/d\lambda = 0$. Thus, the df/dL term can be cancelled from both sides. Thus we arrive at the same equation of motion.

Following the above discussion, we now define our new Lagrangian choosing $f(L) = L^2$ to be

$$L := g_{ij}\,\dot{x}^i\,\dot{x}^j. \tag{4.3.9}$$

Note that, this is done just for saving algebra. Same equations would have emerged from the original Lagrangian $\sqrt{g_{ij}\,\dot{x}^i\,\dot{x}^j}$.

Let us now write down the Lagrange's equation for the geodesic:

$$\frac{d}{d\lambda}\left(g_{ik}\,\dot{x}^k + g_{ij}\,\dot{x}^j\right) = 2\frac{d}{d\lambda}\left(g_{ij}\,\dot{x}^j\right) = g_{jk,i}\,\dot{x}^j\,\dot{x}^k. \tag{4.3.10}$$

Then using the chain rule, $d/d\lambda = (dx^k/d\lambda)d/dx^k$, one can write

$$g_{ij}\,\ddot{x}^j + g_{ij,k}\,\dot{x}^j\,\dot{x}^k = \frac{1}{2}g_{jk,i}\,\dot{x}^j\,\dot{x}^k. \tag{4.3.11}$$

Then using the definition of Γ_{ijk}, symmetry of dummy indices and raising indices by multiplying with g^{mi} on both sides and summing over m, one arrives at the geodesic equation

$$\ddot{x}^i + \Gamma^i_{\ jk}\,\dot{x}^j\,\dot{x}^k = 0. \tag{4.3.12}$$

This is the same equation that we got by parallely transporting the tangent along the curve. Specifically for Euclidean geometry, the geodesics satisfy both the criteria for straight lines, namely, that of (i) parallelly transporting the tangent along the curve and (ii) the property of minimum distance between two points.

A remark is in order. In the above discussion, we have used a special parameter λ marking the geodesics, called the *affine* parameter, to derive the results in the simplest form. It can be shown that such a parameter (or a family of parameters $\lambda' = a\lambda + b$, where a, b are constants) always exists (Eisenhart [1926]). In the above discussion λ is an affine parameter.

Coming back to the problem of ants drawing geodesics, in either method the initial position of the ant is given. The ants could draw infinite number of geodesics through this point. But the geodesics can be uniquely identified in a local neighbourhood,

---

of variation we express the coordinates x^i as functions of the parameter λ (or s), to draw the actual trajectory.

either by the tangent or the end point, which are essentially the Cauchy and Dirichlet type boundary conditions needed for solving the geodesic equations.

Derivation 11 *Show that: Great arcs on a two-sphere are Geodesics.*

Let us consider two points A and B specified on the two-sphere and choose spherical polar coordinates θ, ϕ on the sphere, such that both A and B are on the same longitude ($\phi = $ constant). Since ϕ is the same at A and B, one can parameterise the curve with θ. Then the total path length for a neighboring curve joining A and B is,

$$\Delta s_{AB} = \int_A^B \sqrt{d\theta^2 + \sin^2\theta \, d\phi^2} = \int_{\theta_A}^{\theta_B} d\theta \sqrt{1 + \sin^2\theta \left(\frac{d\phi}{d\theta}\right)^2}. \quad (4.3.13)$$

Clearly this integral is minimum among all neighbouring curves if the second term inside the square root is zero. This means that the minimum distance is obtained for a path for which $\phi = $ constant. This is in fact a great circle.

Derivation 12 *Write down the geodesic equation for a two sphere and read off the Christoffel symbols.*

The Lagrangian for two sphere can be taken as $L = \dot{\theta}^2 + \sin^2\theta \, \dot{\phi}^2$. Hence the geodesic equations are:

$$\ddot{\theta} - \sin\theta \, \cos\theta \, \dot{\phi}^2 = 0, \quad (4.3.14)$$

$$\frac{d}{d\lambda}(2\sin^2\theta \, \dot{\phi}) = 0 \Rightarrow \ddot{\phi} + 2\cot\theta \, \dot{\theta} \, \dot{\phi} = 0. \quad (4.3.15)$$

Comparing with the geodesic equation [Eq. (4.3.12)], only non-zero Christoffel symbols are:

$$\Gamma^\theta_{\phi\phi} = -\sin\theta \, \cos\theta \quad (4.3.16)$$

$$\Gamma^\phi_{\phi\theta} = \Gamma^\phi_{\theta\phi} = \cot\theta, \quad (4.3.17)$$

which matches the results in Derivation 6. There it was easy to spot which Γ_{ijk} are non-zero, but not so for Γ^i_{jk}, which was easy here. Note that one must divide the coefficient of $\dot{\theta} \, \dot{\phi}$ by 2 because there are two terms $\Gamma^\phi_{\theta\phi} = \Gamma^\phi_{\phi\theta}$ in the geodesic equation.

4.4 Locally Flat Coordinate System

We know that it is always possible to select two infinitismally separated points on
a curve and join them by a straight line, which is defined as the tangent to the
curve at that point. Since these two points are infinitesimally close, the line joining
them is in fact a part of the curve. Similarly, in higher dimensions one can find an
infinitesimally small volume, where the space can be considered "flat". That is, in
a small neighbourhood of a point in space, it is possible to set up a local Cartesian
coordinate system.

To validate the above statement we first pick a point P and evaluate quantities at
P. We ask the following question: for a given coordinate system x^i and metric g_{ij}, is
it possible to find a coordinate transformation $x^i \rightarrow x'^i = x'^i(x^j)$, such that, its first
derivatives $g_{ij,k}$ vanish at P? The criterion would imply that the coordinate system
is locally Cartesian at P and would ensure that components of two infinitesimally
separated vectors could be directly compared without the need for any elaborate
procedure of parallel transport, because the Christoffel symbols vanish at that point
in this coordinate system. In relativity, this would mean that the space-time is locally
flat and SR would be valid. This also shows that the choice of Riemannian geometry
is appropriate for accommodating the equivalence principle. Further, can the metric
be made into an identity matrix at P?

The answers to the above questions we asked are in the affirmative. To answer
the second question first, we should be able to find n^2 quantities $\partial x^i/\partial x'^j$ such that
the following $n(n+1)/2$ equations are satisfied:

$$g_{ij} \frac{\partial x^i}{\partial x'^k} \frac{\partial x^j}{\partial x'^l} \bigg|_P = g'_{kl} = \delta_{kl}. \tag{4.4.1}$$

For GR we would replace δ_{kl} with η_{kl} and set $n = 4$. But here we take a general
scenario of n dimensions and also we have assumed the metric to be positive definite.
The signature of the metric does not make any difference to the arguments which
follow.

Since for $n \geq 2$, the number of equations is less than the number of parameters,
this is possible. Not only that, it is possible in different ways, which makes sense.
If one can set up a local Cartesian coordinate system around a point P, then that
system can be rotated around P and it would still remain a local Cartesian system.
In relativity we have $n = 4$ and the "rotations" are the Lorentz transformations—we
have 6 free parameters which represent the 3 boosts and 3 parameters describing
rotations in 3-space—for example the 3 Euler angles. This result follows from the
well known theorems in linear algebra. Observe that g_{ij} is just a symmetric matrix at
the point P and it can be brought into the form of δ_{kl} or η_{kl} by a linear transformation
or more correctly by an infinity of linear transformations connected by the rotation
or the Lorentz group as the case may be.

Now we come to the the next question: can one make $g'_{kl,m}$ zero at P? Taking
derivative of the Eq. (4.4.1) at P, shows that

$$\frac{\partial g'_{kl}}{\partial x'^m} = \frac{\partial}{\partial x'^m}\left(g_{ij}\frac{\partial x^i}{\partial x'^k}\frac{\partial x^j}{\partial x'^l}\right),\tag{4.4.2}$$

where we have omitted to write P to avoid clutter. It is understood that the above quantities are evaluated at P. This introduces $n^2(n+1)/2$ more equations [$n(n+1)/2$ combinations of kl and n values of m] and $n^2(n+1)/2$ extra parameters $\partial^2 x^i/(\partial x'^k \partial x'^l)$, which exactly match! So this is just about possible. In fact, as shown in Landau and Lifshitz [1980], a coordinate transformation of the form

$$x'^i = (x^i - x_P^i) + \frac{1}{2}[\Gamma^i{}_{jk}]_P (x^j - x_P^j)(x^k - x_P^k),\tag{4.4.3}$$

where $[\Gamma^i{}_{jk}]_P$ are the Christoffel symbols at a given point P, and this transformation makes the Christoffel symbols in the new coordinate system at P zero or $[\Gamma'^i{}_{jk}]_P = 0$. This can be verified by using the transformation equation for Christoffel symbols, Eq. (4.2.31), and showing that, $[(\partial^2 x'^m/\partial x^j \partial x^k)(\partial x^i/\partial x'^m)]_P = [\Gamma^i{}_{jk}]_P$. Moreover, since this transformation also has the property that $[\partial x^i/\partial x'^j]_P = \delta^i_j$—if one had already made $g'_{ij} = \delta_{ij}$, then this condition remains untouched.

Proceeding further along the same lines, we can then ask, can we make the second derivative of the metric $g_{ij,kl}$ at P equal to zero? The answer is no. The number of second order equations,

$$\frac{\partial^2 g'_{mn}}{\partial x'^k \partial x'^l} = \frac{\partial^2}{\partial x'^k \partial x'^l}\left(g_{ij}\frac{\partial x^i}{\partial x'^m}\frac{\partial x^j}{\partial x'^n}\right),\tag{4.4.4}$$

is $E(n) = n^2(n+1)^2/4$, while the number of parameters $\partial^3 x^i/(\partial x'^j \partial x'^k \partial x'^l)$ is only $P(n) = n^2(n+1)(n+2)/6$. Subtracting the number of parameters from the number of equations yields,

$$E(n) - P(n) = \frac{n^2(n+1)^2}{4} - \frac{n^2(n+1)(n+2)}{6} = \frac{n^2(n^2-1)}{12}.\tag{4.4.5}$$

So the number of equations is always greater than the number of parameters for $n \geq 2$. For example, in four dimensions, the number of equations is 100 while the number of parameters is 80. So one can make 80 of the $g_{ij,kl}$ zero at P, but 20 still remain. As we will see these are exactly the number of independent components of the Riemann curvature tensor defined later in the text. In fact in n dimensions the number of independent components of the Riemann tensor are exactly $n^2(n^2-1)/12$ which matches with Eq. (4.4.5). When $n = 1$, we have $E(n) = P(n)$, which shows that there is no curvature in 1 dimension. Note that here we are talking about the *intrinsic* curvature of a space. A curve has zero intrinsic curvature. A string placed along the curve can be deformed into a straight line without stretching or squashing (that is keeping distances along the string invariant), and so must have the same curvature as a straight line which is zero. See Schutz [1995] for additional discussion.

The above analysis shows that, in general, with coordinate transformations one cannot make the second derivatives of the metric vanish. In relativity this translates to the fact that, if the space-time is curved, it is impossible to find a *global* coordinate system that is Cartesian, that is, gravitation cannot be removed by transforming to a specific global frame.

4.5 Curvature

Curvature is a key property of a multi-dimensional space. It is an intrinsic property of space, which cannot be removed by coordinate transformations. We will see later, that gravity manifests itself as curvature in space-time, so the measurement of curvature is very important for the understanding of gravitation.

Measurement of curvature is not a trivial task. For a long time humans used to think that the earth is flat—flat like a circular disc! When it comes to Earth, our perception of the "vertical" dimension is much smaller compared to the radius of the earth; in this case we are like very small ants on a football who can only see along the ball's surface. In general, to understand gravity, we need to deal with the curvature of $3 + 1$ dimensional space-time, which is impossible to visualise. One cannot even visualise a three sphere, because our brain cannot perceive more than three dimensions.

We ask the question could an ant draw the conclusion that the surface of a sphere is curved by making measurements entirely within the sphere. The answer is yes. The ant could draw a small circle on the sphere and measure its radius along the surface of the sphere and also its circumference. It will find the circumference to radius ratio is less than 2π. It will find that the ratio falls short by the factor $1 - r^2/6R^2$, upto second order in r/R, where r is the radius of the circle, measured along the surface of the sphere, and R is the radius of the sphere. The quantity $K = 1/R^2$ is known as the Gaussian curvature. For a general surface the formula generalises to $1 - Kr^2/6$. This kind of curvature is called the intrinsic curvature because it has been obtained by making measurements entirely within the surface. The other kind of curvature is the extrinsic curvature when the surface is embedded in a three dimensional space, usually Euclidean space. In GR we are interested in the intrinsic curvature of the 4 dimensional spacetime. The Gaussian curvature which is defined for 2-surfaces is intimately connected with the Riemann curvature tensor.

4.5.1 Riemann-Christoffel Curvature Tensor

A key indicator of curvature is that if one parallel transports a vector over a closed loop, the vector does not return to its original orientation. This can be illustrated by performing the following exercise on a two sphere (See Exercise 4.2). A vector parallely transported around a latitude (except the equator) does not return to itself,

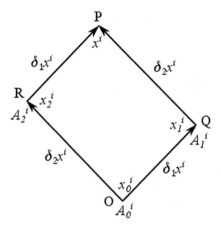

Fig. 4.2 Parallel transport over a closed loop does not bring a vector back to the original orientation. The final difference in orientation, in fact, provides a measure of curvature. A vector $\mathbf{A_0} \equiv A_0^i$ is parallel transported from point O to P via points Q and R. The result is not the same in general—it is path dependent. The difference between the parallely transported vectors measures the curvature, namely, the relevant component of the Riemann tensor

showing that the sphere posseses non-zero curvature or the sphere is a curved space. If one has to measure the curvature at a given point in space, one would construct an infinitesimal closed loop around that point and parallel transport a vector around the loop, the change in the vector is directly related to the measure of curvature.

Let us consider a vector $\mathbf{A_0} \equiv A_0^i$ parallelly transported around the closed rectangle OQPRO, bounded by infinitesimal displacements $\delta_1 x^i$, $\delta_2 x^i$, $-\delta_1 x^i$ and $-\delta_2 x^i$ respectively. The procedure is depicted in Fig. 4.2. To the leading order, this equals the difference between the value of the vector at P if it was transported from O via Q versus via R. That is, parallel transport is path dependent, the transported vector components depend on the path taken.

The difference ΔA^i via two different paths can be calculated. Let us first take path 1 and transport A_0^i from O to Q. We then have:

$$\delta A^i_{O \rightarrow Q} = -\Gamma^i{}_{jk}(O)A_0^j \delta_1 x^k ,$$
$$A^i(Q) = A_0^i + \delta A^i_{O \rightarrow Q} = A_0^i - \Gamma^i{}_{jk}(O)A_0^j \delta_1 x^k , \qquad (4.5.1)$$

where the Christoffel symbols are evaluated at O. Further parallely transport $A^i(Q)$ to P to obtain $A^i(P) \equiv A_1^i$ in a similar way. We then have,

$$A_1^i = A^i(Q) - \Gamma^i{}_{jk}(Q)A^j(Q)\delta_2 x^k , \qquad (4.5.2)$$

where now the Chrisoffel symbols are evaluated at Q. We now Taylor expand the Christoffel symbols to first order around O to obtain their values at Q approximately:

$$\Gamma^i{}_{jk}(Q) = \Gamma^i{}_{jk}(O) + \Gamma^i{}_{jk,l}(O)\delta_1 x^l, \tag{4.5.3}$$

and substitute for $A^i(Q)$ from Eq. (4.5.1) to obtain A^i_1. This substitution gives a rather long expression, but several terms cancel out in the final result. The zeroth and first order terms in δx^k cancel out and the leading order term is of second order. We just write out this second order term explicitly after rearranging dummy indices:

$$[A^i_1]_{\text{second order}} = \left[-\Gamma^i{}_{jk,l} + \Gamma^i{}_{km}\Gamma^m{}_{jl}\right] A^j_0 \delta_1 x^l \delta_2 x^k, \tag{4.5.4}$$

where Christoffel symbols and their derivatives have been evaluated at O.

The same procedure now must be carried out for the path 2 which is $O \longrightarrow R \longrightarrow P$. From this we again obtain a parallely transported vector which we now call A^i_2. We do not need to do a detailed second calculation here because the result can be easily obtained by interchanging $1 \leftrightarrow 2$. We again need to exchange the dummy indices l and k to combine the equations. For the return path we only need to change the sign (we can do this because parallel transport is linear in the differentials) and then we must take the difference. The result is:

$$\Delta A^i = A^i_1 - A^i_2 = R^i{}_{jkl} A^j_0 \delta_1 x^l \delta_2 x^k, \tag{4.5.5}$$

where we have introduced the Riemann-Christoffel curvature tensor:

$$R^i{}_{jkl} := \Gamma^i{}_{jl,k} - \Gamma^i{}_{jk,l} + \Gamma^i{}_{km}\Gamma^m{}_{jl} - \Gamma^i{}_{lm}\Gamma^m{}_{jk}. \tag{4.5.6}$$

We can also get the Riemann tensor by taking limits as the displacements $\delta x^i \longrightarrow 0$ in the above procedure. Then we have the following result. Taking the second covariant derivative of A^i, we have,

$$A^i_{;lk} = (A^i_{;l})_{,k} + \Gamma^i{}_{mk} A^m_{;l} - \Gamma^m{}_{lk} A^i_{;m}$$
$$= (A^i_{,l} + \Gamma^i{}_{ml} A^m)_{,k} + \Gamma^i{}_{mk}(A^m_{,l} + \Gamma^m{}_{nl} A^n) - \Gamma^m{}_{lk}(A^i_{,m} + \Gamma^i{}_{nm} A^n), \tag{4.5.7}$$

We now change the order of the covariant derivatives and take the difference. This difference is essentially the Riemann tensor. Accordingly, we obtain,

$$A^i_{;lk} - A^i_{;kl} = R^i{}_{jkl} A^j. \tag{4.5.8}$$

Since the left-hand side is a difference of two 3rd rank tensors, it is itself a tensor of the 3rd rank and A^i is an arbitrary vector, and therefore by quotient law, $R^i{}_{jkl}$ is a 4th rank tensor.

One can then raise and lower the indices as usual, $R_{ijkl} := g_{im} R^m{}_{jkl}$. In the locally flat coordinate system, since Christoffel symbols vanish, using Eq. (4.2.37) it is easy to see that

$$R_{ijkl} = \Gamma_{ijl,k} - \Gamma_{ijk,l} = \frac{1}{2} (g_{il,jk} - g_{jl,ik} - g_{ik,jl} + g_{jk,il}). \tag{4.5.9}$$

Clearly, R_{ijkl} is symmetric under the interchange of first and second pairs of indices, but antisymmetric within each pair, that is,

$$R_{jikl} = -R_{ijkl}; \quad R_{ijlk} = -R_{ijkl}; \quad R_{klij} = R_{ijkl}. \qquad (4.5.10)$$

It also satisfies the identity,

$$R_{i[jkl]} = \frac{1}{3}(R_{ijkl} + R_{iklj} + R_{iljk}) = 0, \qquad (4.5.11)$$

where we have used Eq. (3.5.13) and $R_{ijlk} = -R_{ijkl}$ to introduce the $R_{i[jkl]}$, the shorthand notation for symmetrisation. Since R_{ijkl} are tensors, all these symmetries are valid in all coordinate systems. The above symmetries imply that the total number of independent components of R_{ijkl} is $n^2(n^2 - 1)/12$, which is 20 for $n = 4$.

Curvature causes distortions in geometry, which are easily visualised and verified for two dimensional surfaces. For example, the sum of three angles of a triangle (a triangle on a surface is formed by joining the 3 vertices by geodesics) is not in general 180°. We already saw that the circumference of a circle differs from 2π times radius. The deviation of these quantities from their standard values shows that the space posseses non-zero curvature. Quantitatively, they are related to the Riemann curvature tensor and the Gaussian curvature.

An extremely important property of curvature is to make geodesics deviate. We discuss this in the next subsection.

4.5.2 Geodesic Deviation

We know that in a plane if two straight lines start parallel to each other, they remain parallel—the perpendicular distance between them remains constant. While if we take two nearby longitudes on a sphere, then at the equator they are parallel, but when produced, meet at the poles—the distance between them decreases. This is the effect of curvature. For the plane the Gaussian curvature is zero, while for the sphere it is $1/R^2$ where R is the radius of the sphere. Also, for the sphere it is positive; this is why the longitudes which are geodesics move closer on the sphere. For a hyperboloid of one sheet, the geodesics starting parallel move away from each other signalling that the Gaussian curvature is negative. Below we will derive an equation for the connecting vector between nearby geodesics and show that the Riemann tensor appears in this equation and is responsible for geodesic deviation. In the context of gravity if we take freely falling nearby particles (they move along geodesics), when the Riemann tensor is non-zero, the particles do not maintain constant distances between them. For example two particles dropped from the same height freely falling towards the centre of the Earth, move closer to each other because their trajectories are directed towards the centre of the Earth. This indicates that the relevant component of the Riemann tensor is positive. From the Newtonian point of view, the particles move closer together because they experience slightly different gravitational accelerations—the difference in the accelerations is what brings them closer—the tidal acceleration, which in the Newtonian picture is the derivative of the force of gravity; the particles are slightly displaced from each other and are in an inhomogeneous gravitational field. The differential of the acceleration due to gravity is described by the Riemann tensor and is the tidal acceleration. Thus curvature can be interpreted in terms of tidal accelerations. This is the force responsible for tides on Earth.

Consider a one parameter family of geodesics parametrised by the parameter μ and with affine parameter λ. Consider nearby geodesics given by $x^i(\lambda, \mu)$ and $x^i(\lambda, \mu + \delta\mu)$. To keep track of the distance between the geodesics we define a connecting vector by expanding in the parameter μ to first order:

$$x^i(\lambda, \mu + \delta\mu) = x^i(\lambda, \mu) + \frac{\partial x^i}{\partial \mu}\delta\mu \equiv x^i(\lambda, \mu) + \eta^i(\lambda, \mu), \qquad (4.5.12)$$

where we have defined the connecting vector:

$$\eta^i = \frac{\partial x^i}{\partial \mu}\delta\mu. \qquad (4.5.13)$$

Note that the partial derivative is evaluated at μ. Since $\delta\mu$ is small, η^i is also small and so we will keep terms in our computations to the first order in $\delta\mu$ or η^i. Then the equations for the two geodesics are:

$$\frac{d^2x^i}{d\lambda^2} + \Gamma^i{}_{jk}\frac{dx^j}{d\lambda}\frac{dx^k}{d\lambda} = 0,$$

$$\frac{d^2x^i}{d\lambda^2} + \Gamma^i{}_{jk}\frac{dx^j}{d\lambda}\frac{dx^k}{d\lambda}$$
$$+ \frac{d^2\eta^i}{d\lambda^2} + \Gamma^i{}_{jk,m}\eta^m\frac{dx^j}{d\lambda}\frac{dx^k}{d\lambda} + 2\Gamma^i{}_{jk}\frac{dx^j}{d\lambda}\frac{d\eta^k}{d\lambda} = 0, \qquad (4.5.14)$$

where we have kept terms to first order in η^i. Subtracting these two equations yields,

$$\frac{d^2\eta^i}{d\lambda^2} + \Gamma^i{}_{jk,m}\eta^m\frac{dx^j}{d\lambda}\frac{dx^k}{d\lambda} + 2\Gamma^i{}_{jk}\frac{dx^j}{d\lambda}\frac{d\eta^k}{d\lambda} = 0. \qquad (4.5.15)$$

To compute the evolution of the connecting vector along the geodesic we consider the intrinsic derivatives of η^i with respect to λ. The first differentiation gives:

$$\frac{D\eta^i}{d\lambda} = \eta^i{}_{;k}\frac{dx^k}{d\lambda} = \left(\frac{\partial\eta^i}{\partial x^k} + \Gamma^i{}_{km}\eta^m\right)\frac{dx^k}{d\lambda} = \frac{d\eta^i}{d\lambda} + \Gamma^i{}_{km}\eta^m\frac{dx^k}{d\lambda}. \qquad (4.5.16)$$

Note that the intrinsic derivative $D\eta^i/d\lambda$ is a contravariant vector. And therefore we can take its second derivative as follows,

$$\frac{D^2\eta^i}{d\lambda^2} = \left(\frac{D\eta^i}{d\lambda}\right)_{;l}\frac{dx^l}{d\lambda}$$

$$= \frac{d^2\eta^i}{d\lambda^2} + \Gamma^i_{km,n}\eta^m\frac{dx^n}{d\lambda}\frac{dx^k}{d\lambda} + \Gamma^i_{km}\eta^m\frac{d^2x^k}{d\lambda^2}$$

$$+ \Gamma^i_{km}\frac{dx^k}{d\lambda}\frac{d\eta^m}{d\lambda} + \Gamma^i_{lp}\frac{dx^l}{d\lambda}\frac{d\eta^p}{d\lambda} + \Gamma^i_{lp}\Gamma^p_{km}\frac{dx^k}{d\lambda}\frac{dx^l}{d\lambda}\eta^m. \quad (4.5.17)$$

Eliminating $d^2\eta^i/d\lambda^2$ from Eqs. (4.5.15) and (4.5.17) we obtain:

$$\frac{D^2\eta^i}{d\lambda^2} = \left[\Gamma^i_{jm,k} - \Gamma^i_{jk,m} + \Gamma^i_{kp}\Gamma^p_{jm} - \Gamma^p_{kj}\Gamma^i_{pm}\right]\frac{dx^j}{d\lambda}\frac{dx^k}{d\lambda}\eta^m$$

$$= R^i_{jkm}\frac{dx^j}{d\lambda}\frac{dx^k}{d\lambda}\eta^m. \quad (4.5.18)$$

We need to qualify the word 'nearby' in the pair of geodesics that we have considered. It is a relative term. What this means is the following. The Riemann tensor is composed of second derivatives of the metric so has the dimensions of inverse length square. Thus the Riemann tensor defines a length scale. The geodesics are nearby if the distance between them characterised by η^i is small compared to this length scale. For example for the sphere, the length scale is decided by its radius R. Therefore nearby geodesics mean the distance between the geodesics is small compared to R.

From the above discussion it can be shown that the necessary and sufficient condition that a manifold admit a global Cartesian system is that $R^i_{jkl} \equiv 0$ (Eisenhart [1926]). A space is called *flat* if the Riemann tensor identically vanishes throughout the space. A space that is not flat is curved; that is, the Riemann tensor is non-zero at least somewhere in the space.

4.5.3 Useful Tensors and Identities

One important identity the curvature tensor satisfies is the Bianchi Identity:

$$R_{ij[kl;m]} = \frac{1}{3}(R_{ijkl;m} + R_{ijlm;k} + R_{ilmk;l}) = 0. \quad (4.5.19)$$

Derivation 13 *Prove the Bianchi identity,*

$$R_{ij[kl;m]} = \frac{1}{3}(R_{ijkl;m} + R_{ijlm;k} + R_{ijmk;l}) = 0.$$

The first equality comes from Eq. (3.5.13) & $R_{ijlk} = -R_{ijkl}$. *Choose locally flat coordinates. In these coordinates,*

$$R_{ijkl,m} = \Gamma^i_{jl,km} - \Gamma^i_{jk,lm}, \ R_{ijlm,k} = \Gamma^i_{jm,lk} - \Gamma^i_{jl,mk}, \ R_{ijmk,l} = \Gamma^i_{jk,ml} - \Gamma^i_{jm,kl},$$
$$(4.5.20)$$

since the first order terms, like $(\Gamma_{ink} \Gamma^n_{jl}),_m = \Gamma_{ink,m} \Gamma^n_{jl} + \Gamma_{ink} \Gamma^n_{jl,m} = 0$, *vanish. Hence, taking the sum after cyclically permuting the indices, we obtain,*

$$R_{ij[kl,m]} = 0.$$

This is however a tensor equation. In a general coordinate system we have,

$$R_{ij[kl;m]} = 0.$$

The Ricci tensor is obtained by contracting the Riemann tensor $R^i{}_{jkl}$ on the first and third index:

$$R_{ij} := R^k{}_{ikj} = \Gamma^k_{ij,k} - \Gamma^k_{ik,j} + \Gamma^k_{mk} \Gamma^m_{ij} - \Gamma^k_{mj} \Gamma^m_{ik} = 2 (\Gamma^k_{i[j,k]} - \Gamma^k_{m[j} \Gamma^m_{k]i}),$$
$$(4.5.21)$$

where the notation follows from Eq. (3.5.11). It is a symmetric tensor.

Further contracting the Ricci tensor gives the Ricci scalar or scalar curvature. The Ricci scalar is defined as below,

$$R := R^i{}_i = g^{ij} R_{ij} = 2 g^{ij} (\Gamma^k_{i[j,k]} - \Gamma^k_{m[j} \Gamma^m_{k]i}).$$
$$(4.5.22)$$

It is a scalar measure of curvature. If the Ricci scalar is positive at a point, a hypersphere of a small radius would enclose smaller volume than that in flat space. Similarly, if the Ricci scalar is negative (saddle like in two dimension) at a point, a hypersphere of the same radius would enclose larger volume than that in flat space.

Derivation 14 *Show that, Ricci tensor is symmetric.*

Choosing locally flat coordinates, we have,

$$R_{ij} := R^k_{\ ikj} = g^{kl} R_{kilj} = \frac{1}{2} g^{kl} \left(g_{kj,il} - g_{ij,kl} - g_{kl,ij} + g_{il,kj} \right). \quad (4.5.23)$$

Clearly, the second and the third terms are symmetric in i and j. The contracted sum of the first and the last term can also be rewritten to ensure the same:

$$\frac{1}{2} g^{kl} (g_{kj,il} + g_{il,kj}) = \frac{1}{2} g^{kl} (g_{lj,ik} + g_{il,kj}) = \frac{1}{2} g^{kl} (g_{jl,ki} + g_{il,kj}). \quad (4.5.24)$$

Since the symmetry of a tensor is independent of the coordinates used, the proof is complete.

The Einstein tensor,

$$G_{ij} := R_{ij} - \frac{1}{2} R g_{ij}, \quad (4.5.25)$$

is also symmetric. The most important property of the Einstein tensor is that it's divergence is zero, which can be proved using the Bianchi identity,

$$G^i_{\ j;i} = 0. \quad (4.5.26)$$

Derivation 15 *Show that, Einstein tensor is divergence-free, $G^i_{\ j;i} = 0$.*

Contracting the Bianchi identity, Eq. (4.5.19), since $g^{ij}_{\ ;k} = 0$, one gets

$$g^{ik} \left(R_{ijkl;m} + R_{ijlm;k} + R_{ijmk;l} \right) = R_{jl;m} + g^{ik} R_{ijlm;k} - R_{jm;l} = 0$$

Contracting the above again to get the divergences

$$g^{jm} \left(R_{jl;m} + g^{ik} R_{ijlm;k} - R_{jm;l} \right) = R^m_{l;m} + R^k_{l;k} - R_{;l} = 2 R^k_{l;k} - R_{,l} = 0.$$

For a scalar, the covariant derivative is the same as the partial derivative and so $R_{;l} = R_{,l}$. From Eq. (4.5.25),

$$G^i_j = g^{ik} G_{jk} = g^{ik} R_{jk} - \frac{1}{2} R g^{ik} g_{jk} = R^i_k - \frac{1}{2} \delta^i_j R.$$

Since, $\delta^k_{l,k} = 0$, $\Rightarrow$ $(\delta^k_l R)_{,k} = R_{,l}$, one has $2R^k_{l;k} - R_{,l} = (2 R^k_l - \delta^k_l R)_{;k} = 2 G^k_{l;k} = 0.$

We now define the Weyl tensor. It is given by:

$$C_{ijkl} := R_{ijkl} - \frac{2}{n-2}(g_{i[k} R_{j]l} - g_{j[k} R_{l]i}) - \frac{2}{(n-1)(n-2)}R\, g_{i[k}\, g_{j]l}\,.$$

(4.5.27)

In 4 dimensions the Weyl tensor has 10 independent components. The Riemann tensor can thought of to be composed of the Ricci tensor and the Weyl tensor, each having 10 independent components adding up to 20 independent components of the Riemann tensor. In Einstein's theory, the gravitational field (curvature) in vacuum is completely described by the Weyl tensor, because the Ricci tensor vanishes in vacuum.

Exercises

1. Compute the Christoffel symbols in polar coordinates on a plane using Eq. (4.2.9). The unprimed coordinates are (r, θ) and the primed ones are the Cartesian coordinates (x, y).

2. *Parallel transporting a vector on a sphere and origami*: Consider a vector $A^i = (A^\theta, A^\phi)$ on a unit sphere with coordinates (θ, ϕ) at a point whose coordinates are $(\theta = \alpha, \phi = 0)$. Parallely transport the vector along the latitude $\theta = \alpha$ from $\phi = 0$ to $\phi = 2\pi$. Show that the vector rotates by the angle $2\pi \cos \alpha$. In general the vector does not come back to its initial direction except when $\theta = \pi/2$, that is, on the equator (the equator is a geodesic). Carry out the following steps:

   (a) Start with the equation $\delta A^i = -\Gamma^i_{\ jk} A^j \delta x^k$ and set $\delta x^k = (0, \delta\phi)$ because $\delta\theta = 0$ along the latitude.

   (b) Obtain two differential equations (first order) for A^θ and A^ϕ by taking the limit $\delta\phi \longrightarrow 0$. They are coupled. Decouple them by differentiating one of them and substituting in the other.

   (c) Fix the initial conditions: $A^\theta = 1$, $A^\phi = 0$ at the starting point $\phi = 0$ and hence solve the differential equations to obtain:

   $$A^\theta = \cos(\phi \cos \alpha), \qquad A^\phi = -\frac{1}{\sin \alpha} \sin(\phi \cos \alpha)$$

   Hence parallely transport A^i from $\phi = 0$ to $\phi = 2\pi$ and show that it rotates by the angle $2\pi \cos \alpha$.

   This result can also be obtained by origami: Fit a cone to the sphere so that it is tangent to it at the latitude $\theta = \alpha$. Remove and cut the cone along its slant height and spread it out flat. Draw a vector along the radial direction on the rim at the right cut. Draw vectors parallel to this vector in flat space all along the rim. Fit the cone back onto the sphere with the vectors. You will see the vector turning as you traverse the rim around the sphere. The circular rim of the flattened cone subtends an angle of $2\pi \cos \alpha$. See Fig. 4.3 below:

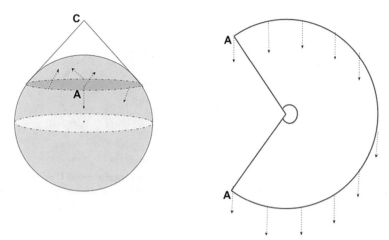

Fig. 4.3 The figure shows how a vector changes its angle with respect to the latitude as it is parallely transported around the latitude in a full circle. We have taken $\alpha < 60°$ so that $2\pi \cos \alpha > \pi$. The vector's initial position is at A, along $\mathbf{e}_\theta$, when it returns to A it points in a different direction. A cone with vertex C is fitted to the latitude, and is cut along its slant side. It spreads into a portion of a disc as shown on the right. If the vector is now transported in flat space, because the disc is flat, one sees that with respect to the tangent to the latitude, it has turned through an angle $2\pi \cos \alpha$. If one now sticks back the cone with the parallely transported vectors back onto the sphere, identical result ensues. *Credits:* Ankit Bhandari

3. Show that in n dimensions the Riemann tensor has $n^2(n^2 - 1)/12$ independent components.

4. **Two Torus:** Consider a two torus of constant radii R and r with $r < R$. Let ϕ and θ be the angle coordinates, where ϕ describes the bigger circle with radius R, and θ corresponds to the smaller circle.

   In Fig. 4.4 the position vector $\mathbf{x}(\theta, \phi) = \mathbf{R}(\phi) + \mathbf{r}(\theta, \phi)$ describes the torus, where,

   $$\mathbf{R}(\phi) = R(\cos \phi, \sin \phi, 0), \quad \mathbf{r}(\theta, \phi) = r \, (\cos \theta \cos \phi, \cos \theta \sin \phi, \sin \theta).$$

   (a) Find the metric tensor $ds^2 = d\mathbf{x} \cdot d\mathbf{x}$. Alternatively, by perusing the Fig. 4.4 and using the fact that the coordinate curves of θ and ϕ are orthogonal, obtain your answer.

   (b) Evaluate the Christoffel symbols and Riemann curvature tensor.

   (c) Show that the curvature is positive on the outer side, negative on the inner side and zero on the upper and lower circles.

5. Geodesic deviation on a unit sphere:

   (a) Show that in spherical coordinates (θ, ϕ) the Riemann tensor component $R^\theta_{\phi\theta\phi} = \sin^2 \theta$.

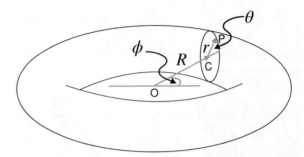

Fig. 4.4 Schematic diagram of a two torus. R is the radius of the big circle with O as origin and the angle ϕ as the azimuthal coordinate. C lies on the circle of radius R. Then with centre C describe another circle (smaller) with radius r in a vertical plane orthogonal to the tangent to the bigger circle. Here θ is another coordinate which marks the point P on that circle. Note that θ is the angle between the vectors **OC** and **CP**

(b) Show that by parallely transporting the vector $A^i = (A^\theta, A^\phi)$ around an infinitismal loop at (θ, ϕ) on the sphere as in Fig. 4.2, show that:

$$\Delta A^\theta = \sin^2 \theta \, A^\phi \, \delta\theta \, \delta\phi.$$

Hence obtain the same result as in part (a).

(c) Consider two nearby longitudes wih ϕ_0 and $\phi_0 + \Delta\phi$, where $\Delta\phi << 1$. Consider the equator $\theta = \pi/2$ and let η^i be the connecting vector connecting the two longitudes (geodesics) at the equator. Then $\eta^\theta = 0$, $\eta^\phi = \Delta\phi$ at the equator. Write down the geodesic deviation equation and choosing the right initial conditions at the equator show that the solution to the equation is $\eta^\phi = \Delta\phi \sin\theta$ and $\eta^\theta = 0$. This independently shows that the longitudes meet at the poles.

6. A coordinate system is called Cartesian if the metric components g_{ij} in this coordinate system are constants. Show that a Riemannian space admits a *global* Cartesian coordinate system if and only if it is flat, that is, the Riemann tensor identically vanishes throughout the space. (Hint: Use the geodesic deviation equation for the "only if" part.)

7. *The topological interpretation of Bianchi identities:* Consider a parallelopiped along the coordinates x^1, x^2, x^3 and parallely transport a vector along all the edges so that looking from outside at the faces of the parallelopiped, the vector is transported in the anticlockwise (positive) direction. Since the vector in this way is transported along each edge once in opposite directions, the total change in the vector is zero, or the final vector is the same as the initial vector. From this observation derive the Bianchi identities. (MTW dramatically remarks "boundary of a boundary is zero".)

Chapter 5
Einstein's Equations

5.1 Einstein's Field Equations

The curvature of spacetime manifests itself as gravitation. But since gravity is caused by matter distribution, one must connect the curvature to the matter distribution. This is exactly what Einstein's equations do. They connect the curvature tensor to a tensor which describes the matter distrbution. But there are several conditions that must be satisfied by the field equations. First of all the equations should be tensor equations so that they are true for all observers or hold in all coordinate systems. Secondly, Einstein's theory must reduce to Newton's theory to a very good approximation in the limit of weak fields and slow motion of sources.

Newton's equation for the Newtonian gravitational potential ϕ is given by:

$$\nabla^2 \phi = 4\pi G\rho \,, \tag{5.1.1}$$

where ρ is the mass density and G the Newton's constant of gravitation $\simeq 6.67 \times 10^{-8}$ in cgs units. Clearly Newton's equation is not Lorentz covariant in SR. This can be seen as follows: for an observer moving with uniform velocity v, the mass transforms as $m \longrightarrow m\gamma$, while the volume shrinks as $V \longrightarrow V/\gamma$ because of Lorentz contraction and thus the $\rho \longrightarrow \rho\gamma^2$. Since there are two factors of γ this indicates that the ρ must be a component of a second rank tensor. Also in SR mass and energy are interconvertible and so ρ or ρc^2 should appear in the equations. Moreover, energy is just one component of the 4-momentum vector and so momentum must also appear in the Einstein's equations and thus must also be a source of gravitation and affect the curvature. The entity that fits the bill is the energy momentum tensor T_{ik}—infact the zero-zero component of T_{ik} is $T_{00} = \rho c^2$. We will physically describe this tensor in the next section. This is the right hand side of Einstein's equations. The left hand side of the equations describe the curvature. The tensor describing the curvature must be of second rank and also symmetric since T_{ik} is of second rank and symmetric. Thus the Riemann tensor by itself cannot be used but its contraction, namely, the

S. Dhurandhar and S. Mitra, *General Relativity and Gravitational Waves*, UNITEXT for Physics, https://doi.org/10.1007/978-3-030-92335-8_5

Ricci tensor R_{ik} is a good candidate. But even this will not do by itself because the energy-momentum tensor T_{ik} has vanishing divergence:

$$T^{ik}_{;k} = 0.$$
(5.1.2)

This equation implies the conservation of energy and monentum. But R^{ik} is not divergence free. But this can be easily remedied by subtracting its divergence. Thus we arrive at the Einstein tensor:

$$G_{ik} = R_{ik} - \frac{1}{2}Rg_{ik},$$
(5.1.3)

where R is the Ricci scalar. By the Bianchi identities, as shown in Eq. (4.5.26) of the last chapter the divergence of this tensor vanishes. Thus we have the Einstein's equations:

$$R_{ik} - \frac{1}{2}Rg_{ik} = \kappa T_{ik},$$
(5.1.4)

where κ is a constant that must be determined. This is achieved by going to the weak field limit and then comparing with Newton's equation Eq. (5.1.1). The energy-momentum tensor can be chosen to have dimensions of either pressure or mass density—one is related to the other by the factor of c^2. We make the former choice. Then the constant κ becomes $8\pi G/c^4$. Thus the Einstein's equations are:

$$G_{ik} = R_{ik} - \frac{1}{2}Rg_{ik} = \frac{8\pi G}{c^4}T_{ik}.$$
(5.1.5)

We will show later in the chapter that in the weak field slow motion limit these equations reduce to the Newton's equation Eq. (5.1.1).

The Einstein tensor G_{ik} can also be obtained via the action principle. The relevant action is called the Einstein-Hilbert action and is given by:

$$\mathfrak{A} = \int_{\mathcal{V}} R \sqrt{-g} \, d^4x$$
(5.1.6)

This action is varied with respect to g^{ik} with the boundary conditions that δg^{ik} and its first derivatives vanish on the boundary of $\mathcal{V}$. This approach is followed in various other books and we refer the interested reader to these (Narlikar 2010). Here we do not further elaborate.

The Einstein's equations are second order coupled nonlinear partial differential equations for the metric components g_{ik}. Since the Riemann tensor involves the second order derivatives of the metric, the equations are also of second order. They are also nonlinear as they involve products of g_{ik} and their first and second order derivatives with respect to the coordinates. The physical meaning of this is that a gravitational field also produces a gravitational field since it carries energy. In contrast, the Newton's equation is linear and so also are the Maxwell's equations of electromagnetism. The implication is that in GR we cannot superpose solutions as we could conveniently do in Newton's theory of gravitation and also in electromagnetism. This makes solutions hard to obtain. Also the equations are coupled. We can make a counting argument about the number of equations: From the symmetry of the Einstein tensor, we have 10 equations (symmetric matrix in 4 dimensions) in the 10 functions g_{ik}. However, the Bianchi identities or vanishing of the divergence of the Einstein tensor, puts 4 conditions on the 10 equations. This brings down the number of independent equations to 6. But this is balanced by the fact that there are 4 degrees of freedom in choice of coordinates. Thus there are really 6 independent g_{ik}. Thus mathematically the problem is well posed.

However, in practice obtaining exact analytic solutions to Einstein's equations is extremely difficult. One must usually impose symmetry conditions so that the problem becomes tractable. But this has the disadvantage of leading us away from realistic astrophysical situations. Nevertheless we have the Schwarzschild solution for a spherically symmetric empty spacetime with a central point mass and the corresponding uncharged, non-spinning black hole. We also have the Kerr solution of the uncharged rotating black hole. There are also solutions for charged rotating black holes—the Kerr-Newman solutions. But even in the simple case of the two body problem, even considering point masses, there is no exact solution in GR—one must stitch together approximate solution (post-Newtonian) to a numerical solution (merger) onto an analytic solution of quasi-normal modes to obtain the full evolution of the orbits. Astrophysically, however, this solution has attained enormous importance with the construction of laser interferometric gravitational wave observatories around the world. The inspiral and the final coalescence of two compact objects such as neutron stars or black holes is among the promising gravitational wave sources for these observatories.

We now describe the energy-momentum tensor which appears on the right hand side of Einstein's equations and describes the matter and energy sources.

5.2 The Energy-Momentum Stress Tensor

The distribution of matter in Einstein's theory is described by the energy-momentum tensor T^{ij}. It is defined to be symmetric. The tensor in its simplest form describes a dust distribution which is characterised by just one function ρ, the density or the equivalently the energy density. Or it can describe what is called a perfect fluid— a fluid which obeys Pascal's law in its rest frame—pressure is isotropic. We will describe the various cases and provide their expressions below. It can also describe even more complex systems. Below we mention the physical meaning of each of its components, with the indices j, k taking values 1, 2, 3 only:

- T^{00} is the energy density ρc^2.
- cT^{0k} is the energy flux across the surface $x^k = $ const.
- T^{kk} is the pressure on the surface $x^k = $ const.

- In general T^{jk} is the flux of the jth component of momentum across the surface $x^k = $ const.. It represents spatial stresses.
- $c^{-1}T^{k0}$ is the kth component of the momentum-density.

In the above description, the indices j, k take values 1, 2, 3.

The above prescription shows that not only mass, but energy, momentum, pressure, elastic stresses all determine the curvature of spacetime or the gravitational field. They are all sources of gravitation. We may contrast this situation with Newton's theory in which it was only the mass density which produced the gravitational field. A simple example will illustrate this. Consider a sphere of uniform density ρ_0 and consider the following two situations: (i) the sphere does not rotate, (ii) the sphere rotates with some constant angular velocity. In Newton's theory there will be no difference in the gravitational field in the two situations, because the distribution of ρ in the two cases will be identical. But in Einstein's theory, it is the energy momentum tensor that is responsible for gravity and it will be different in the two cases. In the second case, there will also be a non-zero momentum flux as distinct from the first, and also other differences and so the energy momentum tensors will differ in the two cases, giving rise to different gravitational fields. For the rotating sphere, one effect is the dragging of inertial frames—we will discuss this later in the context of rotating black holes.

We now give some important examples of energy momentum tensors. The simplest case is that of dust or a dust cloud. Here there is no pressure and no elastic stresses. It is given by the expression:

$$T^{ik} = \rho c^2 u^i u^k .\tag{5.2.1}$$

Here ρ is the density of matter and u^i is the 4-velocity of the matter. Such a tensor is used in cosmology to obtain the Friedmann models of the universe. The Earth too can be considered to be dust. We may expect the pressure to be maximum near the centre of the Earth. It is of the order of $10^{12}-10^{13}$ dynes per square cm, while the average mass density is ~ 5.5 gm/cc. We must compare p/c^2 with ρ and this turns out to be $p/\rho c^2 \sim 10^{-9}$. Thus the pressure can be ignored to compute the gravity due to Earth. The same is true for other components of T^{ik}, except T^{00} which is the only surviving component. Thus this is the only component, namely the mass density, which is responsible for the gravitational field of the Earth just as in Newton's theory. We will show in the next section that Einstein's theory reduces to Newton's theory for weak fields and slow motion. It is therefore advisable to use the simpler theory of Newton to compute the gravitational field of the Earth for the usual day to day situations.

We now consider the case of the perfect fluid. Consider a fluid element and go to its rest frame. In this frame assume Pascal's law that the pressure is the same in all directions. Choose also locally flat coordinates so that the metric tensor upto the first order is η_{ik}, the metric of SR. The energy-momentum tensor in this frame is diagonal and is given by:

$$T^{ik} = \begin{pmatrix} \rho c^2 & 0 & 0 & 0 \\ 0 & p & 0 & 0 \\ 0 & 0 & p & 0 \\ 0 & 0 & 0 & p \end{pmatrix}, \tag{5.2.2}$$

where ρ and p are the density and pressure respectively in the rest frame of the fluid. This tensor can be transformed to a general coordinate frame. In a general frame the same tensor has the form:

$$T^{ik} = (\rho c^2 + p)u^i u^k - pg^{ik}, \tag{5.2.3}$$

where the metric tensor makes an appearance. See Exercise 5.3. It is important to realise that the quantities p and ρ appearing in Eq. (5.2.3) are measured in the rest frame of the fluid. The perfect fluid tensor is used in Friedmann-Robertson-Walker universes. The pressure here is non-zero. But to solve the Einstein's equations with both pressure field and the energy density field one needs another equation, namely, the equation of state $p = p(\rho)$. With this equation the Einstein's equations can be solved to arrive at the metric, at least in principle.

5.3 The Newtonian Limit of Einstein's Equations

Here we will show that in the weak field slow velocity limit the Einstein's equations go over to the Newton's equation. Weak field means that the spacetime in suitable coordinates, which are nearly Cartesian, can be described by the metric:

$$g_{ik} = \eta_{ik} + h_{ik}, \quad |h_{ik}| \ll 1. \tag{5.3.1}$$

So h_{ik} are quantities which are very small compared with unity. We will see how small they are in typical cases later in the section. Secondly, slow velocity means the relevant velocities are small compared with the speed of light c. Here we will take the velocities of the sources as well as those of the test particles small compared to the speed of light.

Given these constraints, the Einstein's equations reduce to just one non-vacuous equation, namely, the one with the 00 component:

$$R_{00} - \frac{1}{2} R \, g_{00} = \kappa T_{00}, \tag{5.3.2}$$

where κ will be determined from Newton's theory. It is easy to show that the Riemann tensor to the first order in h_{ik} is given by:

$$R_{ijkl} = \frac{1}{2} \left(h_{il,jk} + h_{jk,il} - h_{ik,jl} - h_{jl,ik} \right), \tag{5.3.3}$$

where we have dropped terms of $o(h^2)$ and higher. Note that the Γ is of order $o(h)$ so that the product of Γs appearing in the Riemann tensor is of order $o(h^2)$ and hence can be neglected. Comparing with Eq. (4.5.9), we can see that the two forms are similar. The physical situations are however different. Here we have taken the gravitational field to be weak, while in Eq. (4.5.9), the gravitational field need not be weak—we have only chosen a locally flat coordinate system.

In order to obtain the Ricci tensor, we must multiply the above with g^{ik} and contract on i, k. But since we keep terms to first order in h, the g^{ik} can be replaced by η^{ik}. Thus,

$$R_{jl} = \eta^{ik} R_{ijkl}, \tag{5.3.4}$$

and,

$$R_{00} = \frac{1}{2}\eta^{ik} \left(h_{i0,0k} + h_{0k,i0} - h_{ik,00} - h_{00,ik} \right). \tag{5.3.5}$$

We now drop all the time derivatives—slow motion—and arrive at,

$$R_{00} = -\frac{1}{2}\eta^{ik} h_{00,ik}$$
$$\simeq \frac{1}{2}\nabla^2 h_{00}, \tag{5.3.6}$$

where again we have dropped the time derivatives in D'Alembertian to arrive at the Laplacian ∇^2. Contracting the Einstein's equations $R_{ik} - \frac{1}{2}Rg_{ik} = \kappa T_{ik}$ leads to $R = -\kappa T$, where $T = g^{ik}T_{ik}$, the trace of T_{ik}. We can then write Einstein's equations as,

$$R_{ik} = \kappa \left(T_{ik} - \frac{1}{2} g_{ik}T \right). \tag{5.3.7}$$

For the Newtonian limit the T_{ik} has just one non-zero component $T_{00} = \rho c^2$. Thus the trace is also $T = \rho c^2$. These considerations lead to:

$$R_{00} = \frac{1}{2}\kappa\rho c^2. \tag{5.3.8}$$

From Eqs. (5.3.6) and (5.3.8) we obtain:

$$\nabla^2 h_{00} = \kappa\rho c^2. \tag{5.3.9}$$

The next task is to connect the metric with the Newtonian potential ϕ. This either can be achieved through the geodesic equation in the weak field slow motion approximation or through gravitational redshift. In this limit, choosing nearly Cartesian coordinates, the geodesic equation reduces to:

$$\frac{d^2x^i}{dt^2} = -c^2\Gamma^i_{\ 00} \qquad i = 1, 2, 3. \tag{5.3.10}$$

The justification for the above is that for low velocities $ds \simeq cdt$ and on the RHS only the 00 term survives as the other terms are of order v/c or $(v/c)^2$. In the same approximation we also have,

$$\Gamma^i{}_{00} = -\frac{1}{2}\eta^{ik}h_{00,k} = -\frac{1}{2}h^i{}_{00}. \tag{5.3.11}$$

Since we are in Cartesian coordinates, we can pull the index i down and we get,

$$\frac{d^2x^i}{dt^2} = -\frac{1}{2}c^2 h_{00,i}. \tag{5.3.12}$$

This equation can be compared with the Newtonian one,

$$\frac{d^2x^i}{dt^2} = -\frac{\partial\phi}{\partial x^i}. \tag{5.3.13}$$

Comparing the two equations we have:

$$h_{00} = \frac{2\phi}{c^2}. \tag{5.3.14}$$

This correspondence can also be worked out from the gravitational redshift formula. Comparing Eq. (5.3.9), Eq. (5.3.14) and Newton's equation $\nabla^2\phi = 4\pi G\rho$ we determine κ:

$$\kappa = \frac{8\pi G}{c^4}. \tag{5.3.15}$$

We now work out how small are h_{ik}, in particular h_{00} compared to unity in typical situations. Consider the Sun's field at the position of the Earth's orbit. An order of magnitude calculation requires us to compute, ϕ/c^2 where $\phi \sim GM_\odot/R$, $M_\odot \sim 2 \times 10^{33}$ gm is the mass of the Sun and $R \sim 1.5 \times 10^8$ km, the average approximate distance between the Earth and the Sun. The computation gives $h_{00} \sim 10^{-8}$. Since h_{00} is 8 orders of magnitude less than unity, we are justified in considering the field as weak—we can safely drop terms of the order of $o(h^2)$ and higher. Similarly, the h_{00} at the surface of the Earth due to Earth's gravitational field is of the order of 10^{-9}.

Exercises

1. *Perfect fluid energy-momentum tensor*: Consider a small element of a fluid and assume that in the rest frame, which is assumed to be locally flat, the fluid obeys Pascal's law that the pressure is equal in all directions. Let x'^i be the coordinates of this locally flat rest frame, then in this frame, the fluid is described by the energy momentum tensor:

$$T'^{ik} = \begin{pmatrix} \rho c^2 & 0 & 0 & 0 \\ 0 & p & 0 & 0 \\ 0 & 0 & p & 0 \\ 0 & 0 & 0 & p \end{pmatrix}. \tag{5.3.16}$$

Carry out the following steps:

(a) In a general frame x^i, introduce the 4-velocity u^i of the fluid. In the rest frame of the fluid $u'^i = (1, 0, 0, 0)$. Then show that,

$$u^i = \frac{\partial x^i}{\partial x'^0}.$$

(b) Show that (metric in locally flat frame is η_{ik}):

$$g^{ik} = \frac{\partial x^i}{\partial x'^m} \frac{\partial x^i}{\partial x'^n} \eta^{mn} = u^i u^k - \sum_{m=1}^{3} \frac{\partial x^i}{\partial x'^m} \frac{\partial x^k}{\partial x'^m}$$

(c) Hence show that in the general frame:

$$T^{ik} = (\rho c^2 + p)u^i u^k - p g^{ik}.$$

2. Show that $h_{00} = 2\phi_E/c^2 \sim 10^{-9}$ at the surface of the Earth, where ϕ_E is the Newtonian potential of the Earth. For this you will require the values of the mass and radius of the Earth.

3. Use the geodesic deviation equation derived in Chapter 4 to compute the LIF at the surface of the Earth as in Exercise 1 of Chap. 2. Follow the steps as given below:

(a) Write down the geodesic deviation equation Eq. (4.5.18) as:

$$\frac{d^2 \eta^i}{c^2 d\tau^2} = R^i_{jkm} U^j U^k \eta^m,$$

where U^j the 4-velocity of the central particle O and instead of λ we have used the proper time τ.

(b) Approximate $U^j = (1, 0, 0, 0)$ since the velocities are much smaller than the speed of light. Hence only the components R^i_{00m} are required. In the weak field slow motion approximation,

$$R_{i00m} \approx \frac{1}{2} h_{00,im} = \frac{1}{c^2} \frac{\partial^2 \phi}{\partial x^i \, \partial x^m}$$

where $\phi = -GM/R$ the Newtonian gravitational potential of the Earth.

(c) Consider the horizontal direction x and write $\eta^x = \Delta x + \delta x(t)$ for the horizontal separations. Also since motions are slow write $d\tau = dt$ and hence arrive at the equation:

$$\frac{d^2 \delta x}{dt^2} \approx R^x_{00x} \, \Delta x = - \left(\frac{\partial^2 \phi}{\partial x^2} \right)_{surface} \Delta x = - \frac{GM}{R_\oplus^3} \Delta x$$

Similar considerations for the vertical separations Δz and δz lead to:

$$\frac{d^2 \delta z}{dt^2} \approx R^z_{00z} \, \Delta z = - \left(\frac{\partial^2 \phi}{\partial z^2} \right)_{surface} \Delta z = 2 \frac{GM}{R_\oplus^3} \Delta z$$

(d) Integrate the above equations with respect to time from 0 to Δt to obtain δx and δz and obtain the same result as in Eqs. (2.3.2) and (2.3.1). From these expressions, obtain the size of the LIF by setting $\delta x = \delta z = \epsilon$.

Chapter 6
Schwarzschild Solution and Black Holes

6.1 Introduction

In this chapter we will discuss the first exact solution of Einstein's equations. This is the Schwarzschild solution and was given almost immediately after Einstein gave his field equations and the theory of GR. The solution is spherically symmetric and obtained in empty space.The solution also represents a black hole. One of the most interesting features of this solution is the event horizon which we will discuss in some detail and give general arguments to establish its one way nature. Other interesting features also will be discussed. Then we will discuss force free orbits of test particles, both with non-zero rest-mass and zero rest-mass such as photons. These are in fact the so called timelike and null geodesics. We will then describe how a photon is red-shifted as it travels away from the gravitating mass. This then naturally leads to the notion of the infinite-red-shift surface. Finally, we briefly describe the solution for the rotating black hole, namely, the Kerr (1963) solution and discuss some of its interesting features.

6.2 The Schwarzschild Metric

The Schwarzschild spacetime is a spherically symmetric solution of Einstein's vacuum equations—this means the equations are solved in the region in which the energy-momentum tensor T_{ik} vanishes identically—$T_{ik} \equiv 0$. As we will show that the solution is also static—this is not assumed but can be deduced from Einstein's equations. This is the essence of what is called as the Birkoff's theorem, which states that any spherically symmetric vacuum solution of Einstein's equations is a piece of Schwarzschild geometry. Such a spacetime can be realised in vacuum outside any spherically symmetric matter distribution. The matter distribution could also be dynamic, for example, a spherically symmetric collapse of a dust cloud. But as long as spherical symmetry is maintained, the solution in vacuum is still static and a piece

© The Author(s), under exclusive license to Springer Nature Switzerland AG 2022   97
S. Dhurandhar and S. Mitra, *General Relativity and Gravitational Waves*,
UNITEXT for Physics, https://doi.org/10.1007/978-3-030-92335-8_6

of Schwarzschild geometry. It is also possible that the matter has completely collapsed into a singularity, in which case we have the Schwarzschild black hole with the inevitable event horizon.

We will start with a general form of a spherically symmetric metric not necessarily static. Such a metric is given by:

$$ds^2 = A(r,t)dr^2 + B(r,t)cdrdt + C(r,t)c^2dt^2 + D(r,t)(d\theta^2 + \sin^2\theta d\phi^2) ,$$
(6.2.1)

where A, B, C, D are arbitrary functions of r and t (Weinberg (1972)). Because of spherical symmetry, the coordinates can be chosen so that there is only one cross term in $drdt$. It is easy to show that by changing to new coordinates $r'(r,t)$ and $t'(r,t)$, we can eliminate the $drdt$ cross term and also reduce $D(r,t) = -r'^2$. We will assume that this has been done. We now drop the primes on r and t and write the metric in the form:

$$ds^2 = e^\nu c^2 dt^2 - e^\lambda dr^2 - r^2(d\theta^2 + \sin^2\theta d\phi^2) ,$$
(6.2.2)

where ν and λ are functions of r and t. The centre of symmetry is at $r = 0$. The metric is usually written in this form so that far away from the centre, the metric confirms to the usual signature of $(+, -, -, -)$ and moreover, we expect that as $r \longrightarrow \infty$, the metric approaches that of SR or flat spacetime so that the gravitational field decays to zero. Such a solution is called asymptotically flat. This is in fact a boundary condition that we will use when deriving the solution.

The general form of the metric is in terms of ν and λ. Our task is to determine these functions from the Einstein's equations and boundary conditions. For this purpose we need to compute the Einstein tensor and in effect the Ricci tensor which in turn is obtained from the Christoffel symbols. Thus our first job is to compute the Christoffel symbols. One could do this directly from the metric or more efficiently via the geodesic equations obtained from variational principles.

In order to simplify the notation, we write $x^0 = ct, x^1 = r, x^2 = \theta, x^3 = \phi$ and denote derivatives with respect of x^0 and x^1 by dot and prime respectively. The non-zero Christoffel symbols are:

$$\Gamma^0{}_{00} = \frac{1}{2}\dot{\nu}, \quad \Gamma^0{}_{01} = \Gamma^0{}_{10} = \frac{1}{2}\nu', \quad \Gamma^0{}_{11} = \frac{1}{2}\dot{\lambda}e^{(\lambda-\nu)} ,$$

$$\Gamma^1{}_{11} = \frac{1}{2}\lambda', \quad \Gamma^1{}_{01} = \Gamma^1{}_{10} = \frac{1}{2}\dot{\lambda}, \quad \Gamma^1{}_{00} = \frac{1}{2}\nu'e^{(\nu-\lambda)} ,$$

$$\Gamma^1{}_{22} = -re^{-\lambda}, \quad \Gamma^1{}_{33} = -r\sin^2\theta e^{-\lambda} ,$$

$$\Gamma^2{}_{12} = \Gamma^2{}_{21} = \frac{1}{r}, \quad \Gamma^2{}_{33} = -\sin\theta\cos\theta ,$$

$$\Gamma^3{}_{13} = \Gamma^3{}_{31} = \frac{1}{r}, \quad \Gamma^3{}_{23} = \Gamma^3{}_{32} = \cot\theta .$$
(6.2.3)

If the $T_{ik} = 0$, then the vacuum Einstein's equations are $R_{ik} - \frac{1}{2}Rg_{ik} = 0$. Taking the trace of this equation, that is multiplying by g^{ik} and contracting over the indices gives the scalar curvature $R = 0$. This in turn implies that the vacuum Einstein equations are $R_{ik} = 0$. So from the above Christoffel symbols we must compute Ricci tensor components and set them to zero. This procedure will give us differential equations for v and λ which we must solve with boundary conditions to obtain the Schwarzschild metric.

For ready reference let us write down the Ricci tensor in terms of the Christoffel symbols:

$$R_{ik} = \Gamma^l_{ik,l} - \Gamma^l_{li,k} + \Gamma^l_{ik}\Gamma^m_{lm} - \Gamma^m_{il}\Gamma^l_{km}. \tag{6.2.4}$$

From this expression for the Ricci tensor and Eq. (6.2.3) we obtain the expressions for the Ricci tensor components in terms of the functions v and λ. We start with R_{01}:

$$R_{01} = \frac{\dot\lambda}{r} = 0. \tag{6.2.5}$$

This shows that λ is a function of r only and does not depend on t. This fact is very useful in simplifying the rest of the equations. Using the equations $R_{00} = R_{11} = 0$, we obtain the following two equations:

$$2v'' + v'^2 - v'\lambda' + 4\frac{v'}{r} = 0, \tag{6.2.6}$$

$$-\frac{1}{2}v'' + \frac{1}{4}v'\lambda' - \frac{1}{4}v'^2 + \frac{\lambda'}{r} = 0. \tag{6.2.7}$$

Multiplying the second equation by 4 and adding to the first equation gives the result:

$$\lambda' + v' = 0. \tag{6.2.8}$$

Integrating the above equation with respect to r we obtain,

$$\lambda + v = f(t), \tag{6.2.9}$$

where f is some arbitrary function of t. We can use this equation to eliminate v. Because, we have in the metric the term:

$$e^v c^2 dt^2 = e^{f(t)-\lambda(r)}c^2 dt^2. \tag{6.2.10}$$

Defining a new time coordinate by,

$$t' = \int dt\, e^{\frac{1}{2}f(t)}, \tag{6.2.11}$$

we obtain the metric in the form:

$$ds^2 = e^{-\lambda}c^2dt'^2 - e^{\lambda}dr^2 - r^2(d\theta^2 + \sin^2\theta d\phi^2) \,. \qquad (6.2.12)$$

We can again relabel the time coordinate by dropping the prime. From Eq. (6.2.7) we get the following equation for λ:

$$\lambda'' - \lambda'^2 + 2\frac{\lambda'}{r} = 0 \,. \qquad (6.2.13)$$

This equation immediately integrates out to:

$$e^{-\lambda} = A\left(1 + \frac{B}{r}\right), \qquad (6.2.14)$$

where A and B are arbitrary constants to be determined from the boundary conditions. The boundary conditions are as $r \longrightarrow \infty$, the metric should approach the flat Minkowski metric of SR and so $e^{\lambda} \longrightarrow 1$ and at large r the Newtonian limit should be reached. The first boundary condition on Eq. (6.2.14) determines the constant $A = 1$. To determine B we must use the Newtonian limit.

Let M be the total mass of the of the matter distribution or we could also consider a point mass. Then the Newtonian potential outside the matter distribution is $\phi = -GM/r$. But we have already seen that in the Newtonian limit, we have, $h_{00} = 2\phi/c^2$ and so $g_{00} \longrightarrow 1 + 2\phi/c^2 = 1 - 2GM/rc^2$, which determines $B = -2GM/c^2$. Thus the boundary conditions fix the arbitrary constants and then the Schwarzschild solution is given by:

$$ds^2 = \left(1 - \frac{2GM}{c^2r}\right)c^2dt^2 - \left(1 - \frac{2GM}{c^2r}\right)^{-1}dr^2 - r^2(d\theta^2 + \sin^2\theta d\phi^2) \,. \qquad (6.2.15)$$

The M that appears in the solution is the total mass energy .

This is an exact solution of Einstein's equations. We have only considered three components of the Ricci tensor, namely, $\{00\}$, $\{01\}$, $\{11\}$ to obtain the solution. What about the other equations? It can be shown that other equations obtained from the remaining components of R_{ik} do not give anything new.

We will now discuss some features of this solution. One defines the "mass" m in length units as:

$$m = \frac{MG}{c^2} \,. \qquad (6.2.16)$$

As can be checked, m has units of length. Consider the Sun; its mass is $M \sim 2 \times 10^{33}$ gms and hence $m \sim 1.5$ km. What is the significance of this number? We can see from Eq. (6.2.15) that g_{00} tends to zero as r tends to $2m$ and g_{11} becomes infinite. So something strange is occuring at $r = 2m$. It might seem there is some kind of a singularity at this value of r. But on closer inspection, one finds that this is merely because the coordinates become singular; the curvature does not become infinite, and moreover, one can choose a locally inertial frame at this location. The Schwarzschild coordinates are singular at $r = 2m$ and the singularity is called a coordinate singularity. One can in fact construct different coordinates in which the metric is perfectly well behaved at $r = 2m$; the Kruskal-Szekeres coordinates (Kruskal (1960)) solve this problem. We do not discuss these coordinates in this book.

Derivation 16 *Consider the strip of the (x, y) plane $-\pi/2 < x < \pi/2$ and y taking all values from $-\infty$ to ∞. Consider the Euclidean metric $ds^2 = dx^2 + dy^2$ on this strip. Change the x coordinate to $u = \tan x$. Then $-\infty < u < \infty$. The metric in these coordinates is:*

$$ds^2 = (1 + u^2)^{-2} du^2 + dy^2 . \tag{6.2.17}$$

One finds that $g_{uu} \longrightarrow 0$ as $u \longrightarrow \pm\infty$ or as $x \longrightarrow \pm\pi/2$. But there is nothing unusual going on at $x = \pi/2$—it is a part of a plane—nothing could be more benign - it is just the fault of the coordinates. If we go back to (x, y) coordinates there is no problem!

The situation is similar in Schwarzschild coordinates at $r = 2m$.

Although the spacetime is well behaved at $r = 2m$, it has special significance. The surface is called the event horizon. We will elaborate on this when we discuss black holes in the next section. An observer with $r < 2m$ cannot send out signals to an observer with $r > 2m$. The surface $r = 2m$ acts as a horizon, where an observer outside $r = 2m$ cannot receive signals from the region inside $r = 2m$. For the Sun this surface is at $r = 2m \sim 3$km. If the Sun were to collapse to a radius less than 3 km it would become a black hole and the horizon, known as the event horizon would have radius of approximately 3 km!

One can get alternative insight into this by considering a photon being radially sent out by an emitter, say at $r = r_e$, and being observed at $r = r_o$, such that $r_o > r_e > 2m$. We will compute the time taken for the photon emitted at r_e to reach r_o. Since the photon is initially aimed radially outwards, it will continue on a radial trajectory, just by symmetry arguments. So we may disregard θ and ϕ from the propagation equations as they are constant along the trajectory. We have,

$$ds^2 = \left(1 - \frac{2m}{r}\right) c^2 dt^2 - \left(1 - \frac{2m}{r}\right)^{-1} dr^2 \equiv 0 , \tag{6.2.18}$$

since $ds = 0$ along a photon trajectory. From this we get:

$$\frac{d(ct)}{dr} = \pm \left(1 - \frac{2m}{r}\right)^{-1} . \tag{6.2.19}$$

These equations in fact describe the light-cones or since we have projected to 2 dimensions, the light lines. In Fig. 6.1 we only show the future light cones, which lie inside the future light lines. The inside of the light cone consists of future time-like vectors and the boundary consists of future null vectors.

As $r \longrightarrow \infty$, $d(ct)/dr \longrightarrow \pm 1$—the light lines are at 45° indicating that the spacetime is that of SR far away from the central mass. As one moves towards the

Fig. 6.1 The figure shows
the future lightcones in
Schwarzschild space-time
outside the event horizon
$r = 2m$

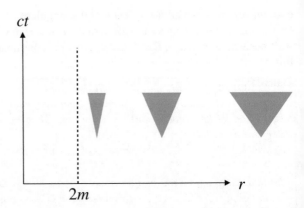

horizon $r = 2m$, the light cones become narrower and finally the two light lines
merge in the coordinate system we are using. For the outgoing light ray one must
choose the positive sign in Eq. (6.2.19), because r must increase as t increases.
Integration of Eq. (6.2.19) within the limits t_e, the time of emission and t_o the time
of observation gives:

$$c(t_o - t_e) = r^\star(r_o) - r^\star(r_e),$$ (6.2.20)

where,

$$r^\star(r) = r + 2m \ln(r - 2m).$$ (6.2.21)

$r^\star$ is called the 'tortoise' coordinate (Misner et al. (1973)). In fact one can easily verify that $r^\star(r_o) - r^\star(r_e) > r_o - r_e$ showing that the photon takes longer to
propagate in Schwarzschild spacetime than in flat spacetime. This is also evident
from the narrowing of the lightcones as one approaches the horizon. In fact as
$r \longrightarrow 2m$, $r^\star(r_e) \longrightarrow -\infty$, implying that it takes longer and longer for the photon
to reach the observer at r_o as the emitter approaches the surface $r = 2m$, until in the
limit the photon is infinitely delayed.

This is just one aspect. One can also prove that the redshift of the photon increases
as the emitter approaches $r = 2m$. So not only does the photon get delayed but also
the "lights are out" as it were.

We will now discuss the event horizon in the general context of black holes and
then specifically for the Schwarzschild blackhole.

6.3 Event Horizons and Black Holes

In this section we generally describe black holes and then show that the Schwarzschild
spacetime where the matter has collapsed beyond $r = 2m$. We investigate the solution
for all $r > 0$ and show that it is a black hole solution. We especially discuss the surface

$r = 2m$ and show that it has special properties. The surface is called the event horizon.

The defining property of the event horizon is its one way nature. It is known generally and popularly that particles, even light cannot escape from a black hole or through the event horizon. Particles from outside the event horizon, in the case of the Schwarzschild black hole the region $r > 2m$, can enter into the region $r < 2m$, but particles inside or on the event horizon cannot escape to the outside or to infinity. Thus the event horizon is a *one way membrane* for particles with non-zero or zero rest mass. Thus one way membranes have a special role to play when defining event horizons or black holes. We will examine this concept in the simplest situation when the spacetime is flat—that of SR. Clearly here there are no black holes, but one-way membranes exist, but they do not surprise us. Why not? This is the question we will attempt to answer. The discussion below is based on very elegant and intuitive arguments given by C. V. Vishveshwara (Vishveshwara (1968)). The lightcone field in the spacetime is of utmost importance in defining one way membranes and black holes.

6.3.1 One Way Membranes in Special Relativity

Let us first look at some examples in SR. The spacetime metric is:

$$ds^2 = c^2dt^2 - (dx^2 + dy^2 + dz^2). \tag{6.3.1}$$

To decide whether a hypersurface—a 3 dimensional subspace of the 4 dimensional spacetime—is a one way membrane, lightcones are of utmost importance—infact they define such a membrane. At any point in the spacetime and more specifically, at each point on the worldline of a particle, we can draw a lightcone. At any point, the tangent to the worldline of the particle, namely, the 4-velocity vector say u^i, lies inside the future light cone. If the particle has non-zero rest-mass then, $u_i u^i = 1$ and $u^0 > 0$. Such a vector is called timelike and future pointing. All such vectors and their positive scalar multiples consititute the inside of the future lightcone—i.e. λu^i, $\lambda > 0$ are the vectors constituting the inside of the future lightcone. On the other hand if the particle has zero rest-mass, then $u_i u^i = 0$ and $u^0 > 0$. This vector is future null and lies on the boundary of the future lightcone. Similarly, we have the past lightcone consisting of past timelike and null vectors with $u^0 < 0$—the null vectors essentially forming a boundary to the set of timelike vectors (since here we have made a reference to boundaries which is a topological notion, we mention that, the relevant topology is the usual topology on $\mathcal{R}^4$). A particle moves from the past lightcone into the future lightcone at any given point on its worldline.

Claim: *The hypersurface $ct = $ const. is a one-way membrane.*

Clearly, when a particle has crossed a given time instant it cannot get back to that instant 'later' on its worldline. This can be easily ascertained by constructing a lightcone at a given space point say (x_0, y_0, z_0) on the hypersurface, say, $ct = $

ct_0. The (future) lightcone one observes lies on *one side* of the hypersurface - we will henceforth consider only future lightcones. Therefore, the particle can only go from $t < t_0$ to $t > t_0$, because the tangent to its worldline must lie inside the future lightcone. But this is true for any point on the hypersurface, since there was no restriction on (x_0, y_0, z_0). So at any given point on $ct = ct_0$, the lightcone lies always on one side of it. Therefore $ct = ct_0$ is a one way membrane. *All one way membranes are defined as those hypersurfaces for which the lightcones lie on one side of the hypersurface.*

Since lightcones are invariant under Lorentz transformations, this statement has an invariant meaning. Every other observer will see this hypersurface as a one way membrane. Thus we can go to any moving frame described by the coordinates (ct', x', y', z') and assert that $ct' = ct'_0$ is also a one-way membrane. But now this hypersurface which is in fact a hyperplane will appear tilted in (ct, x, y, z) coordinates - even then the lightcones all lie on one side of it.

Another example of an hypersurface which is a oneway membrane is the plane wavefront of a zero rest-mass field like the electromagnetic field. If a plane electromagnetic wave travels in the z direction, the surfaces of constant phase are $ct - z = $ const.. In particular $ct - z = 0$ is a one-way membrane. Again the lightcones lie on one side of this hyperplane, but now they touch the hyperplane. Physically this means that once a wavefront has crossed an observer, the observer cannot run and catch up with it. Further example of a one way membrane is a light cone itself, as one can verify by drawing lightcones on its surface. The lightcones again lie on one side of it.

We can characterise a one way membrane in a precise mathematical way through its normals:

All hypersurfaces which have a timelike normal or a null normal are one way membranes.

The $ct = ct_0$ hypersurface has a timelike normal along the ct axis, while the $ct - z = 0$ has a normal $n_i = (1, 0, 0, -1)$ which is null, because $g^{ij} n_i n_j = 0$. A hypersurface which has a timelike normal is called spacelike. This hypersurface has the property that all vectors tangent to it are spacelike. A hypersurface which has a null normal is called a null hypersurface. It has the property that vectors tangent to such a hypersurface are spacelike except for the scalar multiples of the normal vector - here we have the situation that the normal is also a tangent - a consequence of the indefinite scalar product. So the above statement may be reworded.

All spacelike and null hypersurfaces are one way membranes.

The three kinds of hypersurfaces with the lightcones are depicted in Fig. 6.2. The figure on the extreme right shows a *spacelike* hypersurface with light cones on one side. The figure in the middle is a *null* hypersurface where lightcones are also on one side but they touch the hypersurface. Both are one-way membranes. The figure on the extreme left is a *time-like* hypersurface, where the lightcone "cuts" through the hypersurface. This is a two-way membrane and particles can cross from either side.

All the statements made above can be proved rigorously with the help of Schwarz inequality. See Exercise 6.1. Here we will be interested in the null hypersurfaces, because the event horizon is a null hypersurface. But so is $ct - z = 0$ in SR. So

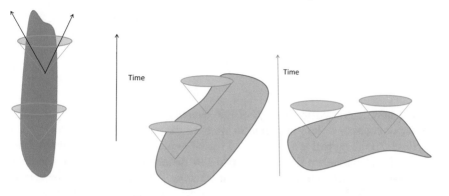

Fig. 6.2 The figures reading from left to right are timelike, null and spacelike hypersurfaces respectively. The null (middle) and spacelike (right) hypersurfaces are one-way membranes. The timelike (left) hypersurface is a two-way membrane

what is different? The difference is that the event horizon $r = 2m$ is a closed and bounded null hypersurface - it does not extend to infinity. This is what makes it special. In SR all one way membranes are infinite—they extend out to infinity; they are non-compact. They are commonplace.

For completeness, we also define a timelike hypersurface which has a spacelike normal. For example, the hypersurface $x = $ const.—the normal is along the x direction and hence spacelike. Now a lightcone, at any point on it, does not lie on one side of this hypersurface—in fact the hpersurface "cuts" through the lightcone. Thus this is a two way membrane—particles can cross the hypersurface from either side.

6.3.2 The Event Horizon in Schwarzschild Spacetime

Here we show that $r = 2m$ is a null hypersurface and therefore a one way membrane. Further we will show that it is the outermost one way membrane.

Since the Schwarzschild solution is spherically symmetric we make the ansatz that the event horizon will also possess the same property. Therefore we will examine the hypersurfaces $r = r_0$, where r_0 is a constant. We will examine the family of such hypersurfaces as r_0 takes different values; we will investigate whether such a surface is timelike, null or spacelike for different values of r_0. For this we need to compute the normal to these hypersurfaces and examine its character, as to whether it is timelike, null or spacelike. For a hypersurface $\psi(x^i) = $ const the normal is given by:

$$n_i = \frac{\partial \psi}{\partial x^i}.$$

(6.3.2)

In our case the $\psi(t, r, \theta, \phi) \equiv r - r_0$ and hence the normal is $n_i = (0, 1, 0, 0)$. The square of its length is obtained by taking $g^{ij} n_i n_j$ which in this case is just:

$$g^{rr} = -\left(1 - \frac{2m}{r_0}\right). \tag{6.3.3}$$

We are only interested in the sign of this quantity which is decided by r_0 being greater or equal to or less than $2m$. So we have the result: when $r_0 > 2m$, g^{rr} is negative and the normal is spacelike and therefore the corresponding hypersurfaces are timelike. This means that they are two way membranes and particles can cross these hypersurfaces from either side. But when $r_0 \leq 2m$, the g^{rr} is either zero or positive and the hypersurfaces are either null or spacelike and therefore one way membranes. In fact $r = 2m$ is a null hypersurface and is the outermost one way membrane and is called the event horizon. Note that this is a closed and bounded hypersurface and does not extend out to infinity as in the case of null hypersurfaces in SR. Such hypersurfaces are called compact. In fact this is the defining property of any event horizon—it is a compact null hypersurface.

6.3.3 Infinite Redshift Surface and the Central Singularity

There are other features of interest of the solution. The hypersurface $r = 2m$ is also an infinite redshift surface. The frequency of the photon emitted by an emitter radially as the emitter approaches $r = 2m$ gets infinitely redshifted when the photon is observed from infinity. We show this later in the text after we have discussed test particle orbits.

Although there is only a coordinate singularity at $r = 2m$, there is a 'real' singularity at $r = 0$; the components of the Riemann tensor become infinite. This means physically that the stresses on a extended body as it approaches $r = 0$ become infinite. However, the components of a tensor becoming infinite also depends on the basis in which it is expressed, it is better to look at an invariant or a scalar constructed out of the curvature tensor which has an invariant meaning independent of the basis. A natural construct is $I = R^{ijkl} R_{ijkl}$—its value turns out to be:

$$I = \frac{48m^2}{r^6}. \tag{6.3.4}$$

This invariant tends to infinity as $r \longrightarrow 0$. Since it is the curvature becoming infinity, it is called a curvature singularity.

6.4 Orbits in Schwarzschild Spacetime

6.4.1 The First Integrals

A test particle—a particle of small mass such that its gravitational field can be safely ignored in the given physical scenario—under no external forces (except gravitation—recall that in GR, gravity is not looked upon as a force) moves along a geodesic of the spacetime. If the particle has non-zero rest-mass then it moves along a timelike geodesic—the tangent to the geodesic at any point $u^i = dx^i/ds$ is a timelike vector, and we have $u^i u_i > 0$—in fact if s is the arclength or c times the proper time then $u^i u_i = 1$. If the test particle has zero rest-mass like a photon, then the tangent to the geodesic is null, that is, if we define the tangent as $u^i = dx^i/d\lambda$ (we cannot use s because $ds = 0$ along the photon trajectory), then $u^i u_i = 0$. We take the parameter λ to be an affine parameter which has been explained in Chapter 4.

Normally, to compute the geodesics, we need to solve the geodesic equations:

$$\frac{d^2 x^i}{d\lambda^2} + \Gamma^i{}_{jk} \frac{dx^j}{d\lambda} \frac{dx^k}{d\lambda} = 0, \qquad (6.4.1)$$

where we have generically used λ. We may set $\lambda = s$ for the timelike geodesics. We need to solve these equations with boundary conditions or initial conditions.

However, for the Schwarzschild spacetime a first principle approach is more useful, because it has several symmetries—it is static, the metric is time independent—and also it is spherically symmetric. Therefore, it makes sense to start from the variational principle and integrate the geodesic equations step by step, by first obtaining the first integrals and identifying the integration constants with physical quantities. We proceed as follows.

First of all, from spherical symmetry, the orbit (geodesic) must be confined to a plane. Because if the particle starts of from a given point and in a given direction, the particle must remain in the plane determined by the, initial space-time point, the tangent vector and $r = 0$—otherwise one hemisphere would be chosen over the other, violating symmetry. We will orient the Schwarzschild coordinates in such a way that this plane is the $\theta = \pi/2$ plane. Thus we need to vary the following integral and obtain its extremum; hence we write:

$$\delta \int (e^\nu c^2 \dot{t}^2 - e^{-\nu} \dot{r}^2 - r^2 \dot{\phi}^2) \, ds = 0, \qquad (6.4.2)$$

where the 'dot' denotes differentiation with respect to s. Thus $\dot{t} = dt/ds$ and so on. We have also considered timelike geodesics first because they are easier to understand. Note that the $\dot{\theta} = 0$, because θ is set to be a constant, namely, $\pi/2$, a consequence of spherical symmetry. Note that this can be rigorously proved from the geodesic equation for θ. If we set $\theta = \pi/2$ and $\dot{\theta} = 0$ in the geodesic equation for θ, we find that $\ddot{\theta} = 0$, establishing that the particle does not leave the $\theta = \pi/2$ plane. See Exercise 3.

We now use the Euler-Lagrange equations to obtain the orbits. Consider a function $F(\dot{x}^i, x^k)$ and we require:

$$\delta \int F(\dot{x}^i, x^k) \, ds = 0, \qquad (6.4.3)$$

where the dot represents d/ds. This condition on the integral is converted into a differential condition by the Euler-Lagrange equations:

$$\frac{d}{ds} \left(\frac{\partial F}{\partial \dot{x}^i} \right) - \frac{\partial F}{\partial x^i} = 0. \qquad (6.4.4)$$

Here we have $F = e^\nu c^2 \dot{t}^2 - e^{-\nu} \dot{r}^2 - r^2 \dot{\phi}^2$ and $x^0 = ct$. If F does not depend on explicitly on x^i for some specific i, then,

$$\frac{d}{ds} \left(\frac{\partial F}{\partial \dot{x}^i} \right) = 0 \Rightarrow \frac{\partial F}{\partial \dot{x}^i} = \text{const.} \equiv p_i, \qquad (6.4.5)$$

where we have defined the conjugate momentum p_i. If F does not explicitly depend on some coordinate x^i, then the momentum conjugate to it, namely, p_i is conserved. We see that in this case, F does not depend explicitly on ct and ϕ and thus p_0 and p_ϕ are conserved. This fact provides us with two first integrals. We already have two other integrals: $\theta = \pi/2$ and since, we must have, $ds^2 = e^\nu c^2 dt^2 - e^{-\nu} dr^2 - r^2 d\phi^2$ along the orbit, dividing by ds^2, we have another first integral:

$$e^\nu c^2 \dot{t}^2 - e^{-\nu} \dot{r}^2 - r^2 \dot{\phi}^2 = 1. \qquad (6.4.6)$$

This equation is just the conservation of 4-momentum of the test particle, which is easily seen by multiplying Eq. (6.4.6) by $m_0^2 c^2$. Then this equation implies $g_{ij} p^i p^j = m_0^2 c^2$, because $p^i = m_0 c \dot{x}^i$ is the 4-momentum.

Since F does not depend on t explicitly, we have,

$$e^\nu \frac{c \, dt}{ds} = E, \qquad (6.4.7)$$

where E is a dimensionless constant—called the constant of motion. We label it as E, because it is essentially the energy of the test particle. Taking the limit as $r \longrightarrow \infty$, we have $e^\nu = 1 - 2m/r \longrightarrow 1$ and $ds^2 \longrightarrow c^2 \, dt^2 - (dx^2 + dy^2 + dz^2)$ and therefore:

$$e^\nu \frac{c \, dt}{ds} \longrightarrow \frac{c \, dt}{\sqrt{c^2 \, dt^2 - (dx^2 + dy^2 + dz^2)}} = \frac{1}{\sqrt{1 - v_\infty^2/c^2}} \equiv E, \qquad (6.4.8)$$

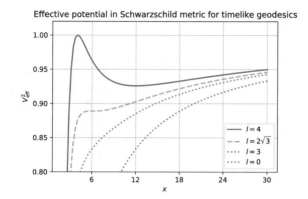

Fig. 6.3 The effective potential in Eq. (6.4.11) is plotted versus $x = r/m$ in figure for various values of $l = L/m$. Essentially two cases arise according as $l > 2\sqrt{3}$ or $l < 2\sqrt{3}$. When $l > 2\sqrt{3}$ the potential has a maximum and particles coming in from infinity with E^2 less than the maximum experience a bounce. If however, E^2 is greater than the maximum, they fall into the black hole. If $l < 2\sqrt{3}$ an ingoing particle always falls into the black hole

where v_∞ is the velocity of the test particle at infinity. But this is just the γ factor at infinity or energy per unit rest-mass of the test particle at infinity divided by c^2. This is the usual result analogous to classical mechanics where the energy is conserved if the Lagrangian does not explicitly depend on time. Note that this interpretation of E tacitly assumes $E \geq 1$, that is the particle can get to infinity. But this is not necessarily the case for bound orbits for which $E < 1$. Then E is just a constant of motion.

In a similar way we obtain the fourth integral of motion since ϕ does not appear explicitly in F. Thus,

$$r^2 \frac{d\phi}{ds} = L\,, \qquad (6.4.9)$$

where cL is the angular momentum of the test particle per unit rest-mass at infinity. Using the above two integrals of motion and Eq.(6.4.6) we solve for $\dot{r}^2$ to obtain,

$$\dot{r}^2 = E^2 - \left(1 - \frac{2m}{r}\right)\left(1 + \frac{L^2}{r^2}\right)\,,$$

$$\equiv E^2 - V_{\text{eff}}^2(r)\,, \qquad (6.4.10)$$

where,

$$V_{\text{eff}}^2(r) = \left(1 - \frac{2m}{r}\right)\left(1 + \frac{L^2}{r^2}\right)\,. \qquad (6.4.11)$$

$V_{\text{eff}}^2(r)$ is called the effective potential. The effective potential is plotted in Fig. 6.3. From the properties of the effective potential we can infer several interesting qualitative results about the orbits.

6.4.2   The Effective Potential

From the expression of the effective potential we see that $V_{\text{eff}}^2 = 0$ at $r = 2m$ and $V_{\text{eff}}^2 \longrightarrow 1$ as $r \longrightarrow \infty$. So in order that the orbit of a particle exist at infinity, we must have $E^2 \geq 1$. The orbit is then said to be unbound. For bound orbits $E^2 < 1$. The 1 represents the rest mass energy of the test particle. To perform a qualitative analysis of the orbits it is useful to examine the extrema of the effective potential. To simplify the algebra, we go over to dimensionless units $x = r/m$ and $l = L/m$. The extrema of the effective potential are obtained by setting:

$$\frac{dV_{\text{eff}}^2}{dx} = 0. \tag{6.4.12}$$

This produces a quadratic in x, $x^2 - l^2 x + 3l^2 = 0$. This equation can be easily solved to yield the roots:

$$x_{\pm} = \frac{l}{2}\left(l \pm \sqrt{l^2 - 12}\right). \tag{6.4.13}$$

From this we see that the roots are real only when $l \geq 2\sqrt{3}$. V_{eff}^2 is plotted in Fig. 6.3 for the cases $l > 2\sqrt{3}$, $l = 2\sqrt{3}$ and $l < 2\sqrt{3}$. When $l > 2\sqrt{3}$, there is a possibility for bound orbits to exist because of the existence of a potential well. But then we also require $E^2 < 1$, but greater than the minimum of V_{eff}^2. When $E^2 > 1$ but less than the maximum value of V_{eff}^2, the particle bounces at the effective potential barrier and returns to infinity. If E^2 is greater than the maximum value of the effective potential, the particle falls into the black hole. For E^2 equal to the extremum values of the effective potential, the particle describes circular orbits - the orbit is stable if the extremum is a minimum and unstable if it is a maximum. When $l = 2\sqrt{3}$, there is a confluence of the extrema and if E^2 is equal to this common value, we again have a circular orbit at $x = l^2/2 = 6$ which means at $r = 6m$. This is the smallest or inner most stable circular orbit in Schwarzschild geometry and from the astrophysical view point very important. When $l < 2\sqrt{3}$, there are no real roots and so there are no extrema for the potential and an ingoing particle does not experience any bounce and falls into the black hole.

In derivation below we describe the Newtonian limit, when the test particle orbits far away from the black hole, that is in the region $r >> 2m$ and when the velocity of the particle is small compared to the speed of light.

The Newtonian Limit of Orbits

Derivation 17 *The Newtonian limit is obtained when particle velocities are small compared to the speed of light, so that the proper time $ds \simeq cdt$, and secondly $r >> 2m$. We first expand Eq. (6.4.10) to obtain:*

$$\dot{r}^2 = E^2 - 1 + \frac{2m}{r} - \frac{L^2}{r^2} + \frac{2mL^2}{r^3}. \qquad (6.4.14)$$

Since $r >> 2m$, the last term is much smaller than the previous term and can be dropped. We put back the G and c and write $L_N = cL$ which is the Newtonian angular momentum per unit mass. We also multiply the above equation by $c^2/2$ and obtain:

$$\frac{1}{2}\dot{r}^2 = \frac{1}{2}(E^2 - 1)c^2 + \frac{GM}{r} - \frac{L_N^2}{2r^2}, \qquad (6.4.15)$$

where now the "dot" represents differentiation with respect to the time t. In the Newtonian picture,

$$E_N = T + V = \left[\frac{1}{2}\dot{r}^2 + \frac{1}{2}r^2\dot{\phi}^2\right] - \frac{GM}{r}, \qquad (6.4.16)$$

where T and V represent the kinetic and potential energies (per unit mass) respectively and E_N is the total Newtonian energy per unit mass. Note that the term in square brackets on the RHS is the kinetic energy for a particle of unit mass moving in a plane.

Observe that Eqs. (6.4.15) and (6.4.16) are identical if we identify E_N with $c^2(E^2 - 1)/2$ and further write $L_N = r^2\dot{\phi}$. With these substitutions we get the equation in the standard form,

$$\frac{1}{2}\dot{r}^2 = E_N - V_N(r), \qquad (6.4.17)$$

where we have defined the Newtonian effective potential as,

$$V_N(r) = -\frac{GM}{r} + \frac{L_N^2}{2r^2}. \qquad (6.4.18)$$

In fact we have $E \simeq 1 + E_N/c^2 \simeq 1$. The 1 as mentioned before represents the rest-mass energy. Note that the constant E as defined is dimensionless - to obtain dimensions of energy, it must be multiplied by the rest-mass times c^2.

6.4.3 Photon Orbits

Again the same argument of spherical symmetry for the orbit, holds for the zero rest mass particle as for the one with non-zero rest-mass—the orbit lies in a plane passing through $r = 0$. We again choose this plane to be $\theta = \pi/2$. This gives us one integral. We again must extremise the functional and we write:

$$\delta \int (e^\nu\, c^2\, \dot{t}^2 - e^{-\nu}\, \dot{r}^2 - r^2\, \dot{\phi}^2)\, d\lambda = 0\,, \tag{6.4.19}$$

but now the dot represents differentiation with respect to a parameter λ. Such a parameter called a affine parameter always exists See Eisenhart (1926) for a more detailed discussion. We again have similar first integrals with s replaced by λ:

$$e^\nu\, c\, \dot{t} = E\,,$$
$$r^2\, \dot{\phi} = L\,,$$
$$e^\nu c^2\, \dot{t}^2 - e^{-\nu}\, \dot{r}^2 - r^2\, \dot{\phi}^2 = 0\,. \tag{6.4.20}$$

where now instead of 1 we have zero in the last of the equations, because the photon has zero rest-mass. Equivalently for a photon, we must have $ds = 0$ - it is a null worldline. This fact reduces one degree of freedom because the tangent to the photon worldline must lie on a null cone. Thus the orbits are not characterised by two independent constants of motion E and L, but their ratio $b = L/E$ suffices. b is called the impact parameter. Eliminating E and L, we write the equation for the orbit as:

$$\frac{1}{r^4}\left(\frac{dr}{d\phi}\right)^2 = \frac{1}{b^2} - \frac{1}{r^2}\left(1 - \frac{2m}{r}\right)\,,$$
$$\equiv \frac{1}{b^2} - B^2(r)\,, \tag{6.4.21}$$

where the potential $B^2(r)$ is defined by the above equation. Below in Fig. 6.4 we plot B^2 as a function of r. The potential has a maximum at $r = 3m$ whose value is $1/27m^2$ (we have plotted $m^2 B^2(r)$ in Fig. 6.4, so the maximum is $1/27$). This potential must be compared with $1/b^2$ for qualitatively describing the orbits. We have a unstable circular orbit of the photon at $r = 3m$ and the value of the impact parameter is $b = 3\sqrt{3}m$. When $b < 3\sqrt{3}m$, the photon does not experience a bounce—an in going photon falls into the black hole, while an outgoing one emitted between $2m$ and $3m$ escapes to infinity. If $b > 3\sqrt{3}m$, then an incoming photon from infinity experiences a bounce from the potential barrier and returns to infinity.

We can write a second order differential equation for the photon orbit in terms of the parameter $u = 1/r$. In terms of this parameter, Eq. (6.4.21) becomes:

Fig. 6.4 We have plotted $m^2 B^2(r)$ versus $x = r/m$ where $B^2(r)$ is given in Eq. (6.4.21). The potential $B^2(r)$ has a maximum at $r = 3m$ of magnitude $1/27m^2$

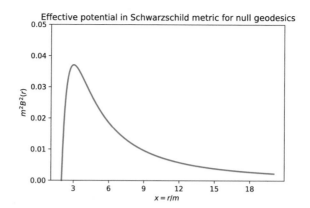

$$\left(\frac{du}{d\phi}\right)^2 = \frac{1}{b^2} - u^2 (1 - 2mu) . \tag{6.4.22}$$

Although of first order, this equation leads to elliptic functions because of the cubic on the RHS. However, this equation can be further differentiated to obtain a somewhat simpler equation although of higher order:

$$\frac{d^2u}{d\phi^2} + u = 3mu^2 . \tag{6.4.23}$$

This form of the equation is useful when discussing the bending of light by a central mass such as the Sun and can be solved perturbatively. We do this calculation in Chapter 7.

6.5   Gravitational Redshift and the Infinite Redshift Surface

Consider a photon emitter at $r = r_e$ and θ, ϕ constant and let it shoot out photons radially outwards towards infinity. Let an observer be at $r = r_o$ with the same θ, ϕ coordinates with $r_o > r_e$. Such an observer or emitter is called static. The situation is elucidated in Fig. 6.5. The photon trajectory follows an outward null radial direction. The slope of the worldline is steeper nearer the horizon because the lightcones narrow down as seen in Fig. 6.1.

Our goal is to relate the energy or frequency of the emitted photons to those of the received photons. Let the 4-velocity of the emitter be u^i. Then we must have $u^i u_i = 1$. Let the wave vector of the photon be k^i. Then we claim that the energy measured by the observer with 4-velocity u^i is just proportional to the scalar product $k^i u_i$, namely, $\hbar k_i u^i$. This can be easily seen by going to the local rest frame of the

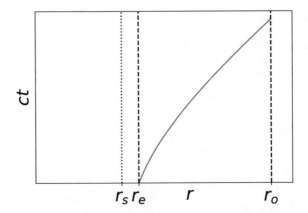

Fig. 6.5 Photons are emitted at $r = r_e$ and propagate radially to the observer at $r = r_o$. The slope of the photon trajectory monotonically decreases until it tends to 45° as $r \longrightarrow \infty$ where away from the horizon, the spacetime approaches flatness. The worldlines of the emitter and observer are shown as vertical dashed lines while the horizon $r = r_s = 2m$ is shown as a vertical dotted line

observer in whose frame $u^i = (1, 0, 0, 0)$ and $k_0 = \omega$ so that $k_i u^i = \omega$ and hence the energy is $\hbar\omega$.

We now need to compute this scalar product $k^i u_i$ at the emitter and at the observer and then take their ratio. We have for the static observer $ds^2 = e^\nu dt^2$, so that,

$$u^i = (e^{-\nu/2}, 0, 0, 0) = \left[\left(1 - \frac{2m}{r} \right)^{-\frac{1}{2}}, 0, 0, 0 \right]. \qquad (6.5.1)$$

We only need to insert the values of r at the emitter and observer to get the corresponding u^i at both locations. Let us denote the redshift by $1 + z$, then,

$$1 + z = \frac{(k_i u^i)_e}{(k_i u^i)_o} = \frac{(e^{\nu/2}k^0)_e}{(e^{\nu/2}k^0)_o}. \qquad (6.5.2)$$

The RHS of the above equation needs to be computed. For this purpose, we need to know how k^0 varies along the null geodesic connecting the emitter to the observer. The first of the energy integrals Eq. (6.4.20), namely, $e^\nu k^0 = $ const. provides the answer. We then get the final result as:

$$1 + z = \frac{(e^{-\nu/2})_e}{(e^{-\nu/2})_o} = \left(\frac{1 - \frac{2m}{r_o}}{1 - \frac{2m}{r_e}} \right)^{\frac{1}{2}} \equiv \left[\frac{(g_{00})_o}{(g_{00})_e} \right]^{\frac{1}{2}}. \qquad (6.5.3)$$

We can see that since we have taken $r_o > r_e$ the photon is redshifted as it moves outwards towards the observer. Also we have written the result more generally in terms of g_{00} the time-time component of the metric tensor which in this form applies to more general situations.

As $r_e \longrightarrow 2m$, the redshift becomes infinite. Therefore the surface $r = 2m$ is also called the infinite redshift surface. We observe that this surface coincides with the event horizon. This happens only for the Schwarzschild solution. In general, for example, for rotating black holes, the two surfaces are different. For the rotating black hole, the infinite red-shift surface is still given by $g_{00} = 0$ but it is not the event horizon. We will discuss the rotating black hole in the Sect. 6.6.

One important case is when the observer is located very far from the horizon, that is in the limit $r_o \longrightarrow \infty$, the redshift is given by $1 + z = (1 - 2m/r_e)^{-1/2}$. In the Newtonian limit when both $r_e, r_o >> m$, and $r_o = r_e + h$ we easily obtain the formula by Taylor expanding the expression and by keeping only the dominant terms:

$$1 + z = 1 + \frac{gh}{c^2},$$

where g is the acceleration due to gravity equal to GM/r_e^2.

In the next chapter we will describe one of the classical tests of GR in which measuring the redshift due the Earth's gravity was one of the tests.

6.6 Kerr Spacetime and the Rotating Black Hole

6.6.1 The Metric and the Horizon

From the astrophysical point of view, a purely non-rotating black hole would be unlikely. Generically, a black-hole would possess spin or non-zero angular momentum. The black holes found in the past few years by the gravitational wave observatories LIGO and Virgo are all in general, spinning having non-negligible angular momentum. Only if the angular momentum is found to be small, then the Schwarzschild solution would approximately be applicable. In this section we briefly describe the rotating black hole which is embodied in the Kerr solution (1963) which describes the spacetime outside a spinning mass. This is perhaps the most important exact solution of Einstein's equation apart from the Schwarzschild.

We write down the metric for a mass M with angular momentum J in Boyer-Lindquist coordinates:

$$ds^2 = \left(1 - \frac{2mr}{\Sigma}\right) c^2 dt^2 - \frac{\Sigma}{\Delta} dr^2 - \Sigma d\theta^2 - \left(r^2 + a^2 + \frac{2ma^2 r \sin^2 \theta}{\Sigma}\right) \sin^2 \theta \, d\phi^2$$

$$+ \frac{4mra \sin^2 \theta}{\Sigma} c \, dt \, d\phi, \tag{6.6.1}$$

where $a = J/cM$ is the angular momentum per unit mass of the black hole and $m = GM/c^2$. Both a and m have dimensions of length. The quantities Δ and Σ are defined by,

$$\Delta = r^2 - 2mr + a^2 , \qquad \Sigma = r^2 + a^2 \cos^2 \theta . \qquad (6.6.2)$$

This is an axially symmetric solution. It is not spherically symmmetric like the Schwarzschild solution. If we put $a = 0$ in the Kerr metric described by Eqs. (6.6.1) and (6.6.2), then the metric reduces to the Schwarzschild metric.

The horizon is given by the root of the equation $\Delta = 0$. Since this is a quadratic, it has two roots $r_\pm$. The larger root $r_+ = m + \sqrt{m^2 - a^2}$ defines the horizon. The other root $r = r_-$ also defines a horizon but since $r_- < r_+$ it lies inside the surface $r = r_+$. A particle coming from large r first encounters $r = r_+$. When $r < r_+$, $\Delta < 0$ and g_{rr} changes sign and the vector in the radial direction becomes timelike. Thus a particle entering $r = r_+$ from outside cannot return to the outside and escape to infinity. Thus $r = r_+$ is the event horizon.

6.6.2   The Static Limit, Frame Dragging and the Ergosphere

We saw that in the case of the Schwarzschild solution, a particle or observer can remain at a fixed (r, θ, ϕ) only outside the event horizon, that is, when $r > 2m$. These are called *static* observers. Now we ask the same question for the Kerr solution. An observer can remain at a fixed (r, θ, ϕ) only when $g_{00} > 0$. Note that the subscript 0 represents $x^0 = ct$. Thus the limit for such observers is obtained when $g_{00} = 0$ or when:

$$r_{\text{static}}(\theta) = m + \sqrt{m^2 - a^2 \cos^2 \theta} . \qquad (6.6.3)$$

The above equation says that static observers only exist outside the surface $r > r_{\text{static}}(\theta)$. What happens inside this surface? Note that the event horizon lies inside the static limit—the static limit surface only touches the event horizon at the poles $\theta = 0, \pi$. In the region $r_+ < r < r_{\text{static}}(\theta)$ the observers must rotate along with the black hole - they need not fall into the black hole - they can escape to infinity if they wish. This region is called the *ergosphere*. In the Fig. 6.6 the ergosphere is depicted as the shaded dark region. The boundary of the lightly shaded circular region lying inside is the horizon $r = r_+$. If we let go a particle initially at rest from infinity (such a particle has zero angular momentum), as it falls towards the black hole it acquires an angular velocity in the direction of the spin of the black hole and spirals into the black hole. This is seen from the discussion below. Observe that the Kerr metric given in Eq. (6.6.1) is independent of the coordinate ϕ. Then from the Euler-Lagrange equations one obtains:

$$g_{0\phi}c\dot{t} + g_{\phi\phi}\dot{\phi} = \text{const.} , \qquad (6.6.4)$$

where the dot represents derivative with respect to the proper time τ (note we have absorbed the factor of c so that we have the right dimensions for the constant) and

Fig. 6.6 The ergosphere is shown as the dark shaded region. The boundary of the lightly shaded circular region lying inside is the event horizon $r = r_+$

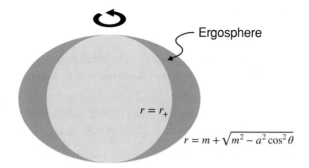

also we have not explicitly written the metric. By taking limits as $r \longrightarrow \infty$, we find $g_{0\phi} \longrightarrow 0$ while $g_{\phi\phi} \longrightarrow -r^2 \sin^2 \theta$ and hence $g_{0\phi} \, c\dot{t} + g_{\phi\phi} \, \dot{\phi} \longrightarrow -r^2 \sin^2 \theta \, \dot{\phi} \equiv -L$, where L is the angular momentum per unit mass of the test particle at infinity. Thus the constant is $-L$ in Eq. (6.6.4). A particle starting from rest from infinity has $L = 0$ (why?) and thus:

$$\frac{d\phi}{dt} = \frac{\dot{\phi}}{\dot{t}} = -\frac{c g_{0\phi}}{g_{\phi\phi}} = \omega(r, \theta) > 0. \qquad (6.6.5)$$

Therefore, such a particle as it falls into the black hole must *co-rotate* with the black hole. In fact if an observer has angular velocity $d\phi/dt = \omega(r, \theta)$ as given in Eq. (6.6.5), then such observers are called *zero angular momentum observers* or ZAMOs. They are also called *locally non-rotating observers* or LNROs. Such observers play an important role from the physical point of view. See Misner et al. (1973). All these effects go by the name of *frame dragging*.

Lastly, as Penrose argued, that the ergosphere can be used for energy extraction from the black hole. Consider the tangent vector ξ to the coordinate $x^0 = ct$ in this region. It is spacelike because $\xi \cdot \xi = g_{00} < 0$. There exist time-like vectors whose scalar product with ξ is negative (see Exercise 1). Thus, one can conceive of a momentum vector $\mathbf{p}$ (future time-like) whose scalar product with ξ is negative, $\mathbf{p} \cdot \xi < 0$. But this is just the energy at infinity $\mathcal{E}$ (Misner et al. (1973)). Therefore, one can have particles inside the ergosphere with momentum $\mathbf{p}$ such that $\mathbf{p} \cdot \xi = \mathcal{E} < 0$. We can then arrange for a particle with energy $\mathcal{E}$ to come in from infinity and split into two particles inside the ergosphere with one of them having $\mathcal{E}_2 < 0$. Because energy is conserved, we have $\mathcal{E} = \mathcal{E}_1 + \mathcal{E}_2$, and so we have $\mathcal{E}_1 > \mathcal{E}$. This particle escapes to infinity with larger energy $\mathcal{E}_1$ than the incoming particle which had energy $\mathcal{E}$. Thus we see that we can extract energy from a rotating black hole. This energy comes at the expense of the black hole losing rotational energy, because the particle with energy $\mathcal{E}_2 < 0$ falls into the hole reducing its energy.

Exercises

1. Use the Schwarz inequality in Euclidean 3-space with scalar product $\mathbf{x} \cdot \mathbf{y} = x_1 y_1 + x_2 y_2 + x_3 y_3$ and norm $|\mathbf{x}| = \sqrt{\mathbf{x} \cdot \mathbf{x}}$ to show that:

   (a) All vectors orthogonal to a time-like vector must be space-like.
   (b) All vectors orthogonal to a null vector must be space-like except multiples of the null vector itself. The null vector is orthogonal to itself.

   (Hint to part (a): Let t^μ be a time-like vector: $|t^0| > |\mathbf{t}|$ and s^μ another vector such that $s^\mu t_\mu = 0$ or $s^0 t^0 = \mathbf{s} \cdot \mathbf{t}$, then from Schwarz inequality in Euclidean space $|s^0 t^0| \le |\mathbf{s}||\mathbf{t}| \Longrightarrow |s^0| < |\mathbf{s}|$. $\mathbf{s}$ is spacelike. A similar argument applies to part (b))

2. Using the coordinate transformation, $v = ct + r^*$, where r^* is the 'tortoise' coordinate, show that Schwarzschild metric can be written as,

   $$ds^2 = \left(1 - \frac{2m}{r}\right) dv^2 - 2\,dv\,dr - r^2 d\Omega^2.$$

3. Show that by using the geodesic equation for the θ coordinate, that if at an initial instant for a test particle $\theta = \pi/2$ and $\dot{\theta} = 0$, then $\ddot{\theta} = 0$, so that the test particle remains confined to the $\theta = \pi/2$ plane.

4. **Kretschmann invariant:** Using the non-zero components of the Riemann tensor for the Schwarzchild metric, compute the Kretschmann invariant $I = R^{iklm} R_{iklm}$ and show that it is $48m^2/r^6$. From this result conclude that the singularity of space-time is at $r = 0$ and not at the event horizon.

5. **Observer falling radially into a black hole:** From the geodesic equation in Schwarzschild metric for an observer falling radially into a black hole of mass M, show the following:

   (a) If the observer initially starts from rest with the radial coordinate r_0, then
   $$\dot{r} = -\sqrt{\frac{2m}{r_0}\left(\frac{r_0}{r} - 1\right)}, \text{ where } m = GM/c^2.$$

   (b) Calculate the proper time taken by the observer, that is, the time measured by his/her clock, to reach $r = 0$. It is finite! (The proper time taken to reach the singularity $r = 0$ is $\pi \left(\sqrt{r_0^3/2m}\right)/c$.)

   (c) However, by calculating dr/dt near the horizon, show that the coordinate time taken for the observer to reach the Schwarzschild radius $r = 2m$ would be infinite.

   (d) Hence argue that no stationary observer outside the Schwarzschild radius would see the falling observer reach the event horizon.

6. **Estimation of tidal force at the event horizon:** Using the symmetries of the Riemann tensor and considering only radial motion, show that:

   (a) In the Schwarzschild geometry the geodesic deviation equation for the radial separation is given by,

$$\frac{D^2}{D\lambda^2}\xi^r = R^r_{ttr}\frac{dx^t}{d\lambda}\frac{dx^t}{d\lambda}\xi^r .$$

(b) The Riemann tensor for this metric is $R_{rttr} = -(2GM/c^2)/r^3$ and it retains its form in the local freely falling coordinate system. Show that the tidal force drops as $1/r^3$.

(c) The tidal acceleration is given by $a = (2GM/r^3)\xi^r$. If tidal force of $100g$ is enough to "sphagettify" a human body of size ~ 1 m, make a rough estimate of the minimum mass of the black-hole whose event horizon could be crossed by a human being falling radially.

(d) How does the radial tidal force at the event horizon compare, for a 10 solar mass black hole with that of a 10^7 solar mass Super Massive Black Hole (SMBH)?

Chapter 7
Classical Tests of General Relativity

7.1 Introduction

General Relativity is a profoundly beautiful theory! Nevertheless, it is a physical theory and any physical theory must stand up to experimental scrutiny. GR is no exception. However in the case of GR or any other theory of gravity there is a fundamental difficulty in designing experiments. Gravity is a weak force - it is the weakest among all the four known forces. In spite of this problem experiments have been designed to test GR but they usually require high precision and hence advanced technology. The three classical tests of GR described below came in the time frame of years to decades, even though the theoretical predictions were available almost since the inception of the theory. A more recent test proposed by Irwin Shapiro in 1964 had to wait for a few years for implementation for the radar technology to improve. The test was based on the precise measurement of radar echo delay from an inner planet. The most recent test and arguably the most advanced one, that probes GR in the strong field limit (unlike the classical tests for which the field is weak), the observation of gravitational waves, took 100 years after they were theoretically predicted.

Altogether GR has been subjected to several experimental tests and it has come out in flying colours. Apart from these experimental tests which reinforce our confidence in general relativity, we have now a modern test of SR and GR, namely, the GPS. The SR time dilation correction and the GR red-shift correction are vetted every minute in our day to day life, without which the whole GPS navigation system, which serves as a backbone of modern life, would miserably collapse [See Exercise 1]. In this chapter we describe the three classical tests that have been known for several decades and further give the calculations for Shapiro delay which is also sometimes referred to as the fourth classical test of GR. The tests as named are listed below:

1. The deflection of light by a central mass;
2. Perihelion shift of Mercury's orbit;
3. Gravitational red-shift;
4. Shapiro delay.

© The Author(s), under exclusive license to Springer Nature Switzerland AG 2022 121
S. Dhurandhar and S. Mitra, *General Relativity and Gravitational Waves*,
UNITEXT for Physics, https://doi.org/10.1007/978-3-030-92335-8_7

The strong field test from the detection of the binary black holes has been covered in later chapters. Gravitational waves, their detection and their data analysis have been discussed in these chapters.

7.2 The Deflection of Light by a Central Mass

Light is bent in a gravitational field. It does not appear to travel in straight lines in the presence of a gravitational field. Let us consider a central mass with mass M having in its exterior a spherically symmetric gravitational field. Then this gravitational field is represented by the Schwarzschild metric. We will consider the field to be weak. Typically we have in mind the problem of a light ray grazing the surface of the Sun. This is the first of the classical tests of general relativity and was historically an important triumph for general relativity. Expeditions were led by no less a figure than Eddington for testing this prediction of GR during an solar eclipse in 1919. The results of the tests bore out the predictions of GR extremely well—the deflection angle was found to be 1.75 arc seconds in accordance with the predictions of GR. We will now proceed to obtain this result below.

We start with Eq. (6.4.23) from the last chapter. We rewrite this equation here for convenience below:

$$\frac{d^2u}{d\phi^2} + u = 3mu^2 . \tag{7.2.1}$$

This equation is nonlinear because of the $3mu^2$ term and therefore not easy to solve exactly. But we do not require the exact solution; an approximate solution suffices, because the nonlinear term is "small" in a sense we will see soon. Consider a light ray which starts from a large distance from the central mass and grazes past the central mass. We choose coordinates so that the mass is at the origin O; the light ray starts from $x \longrightarrow -\infty$ almost parallel to the x-axis with an impact parameter b and goes past the origin towards $x \longrightarrow \infty$. The situation is shown in Fig. 7.1. If the mass were absent, the light ray would travel along a straight line $y = b$ from $-\infty$ to ∞ parallel to the x-axis, if it had started parallel to the x-axis. In polar coordinates this

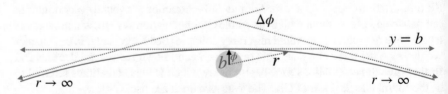

Fig. 7.1 A light ray travels from $x \longrightarrow -\infty$ towards a central mass M located at the origin with impact parameter b and continues towards $x \longrightarrow \infty$. The light ray is deflected by the mass M near the origin. The light ray suffers a deflection of $\Delta\phi \sim 4m/b$. The deflection angle $\Delta\phi$ is grossly exaggerated in the figure for clarity

equation becomes $r \sin \phi = b$. But because there is a central mass, the light ray, as it approaches the origin will bend towards the mass and it will continue almost in a straight line in a slightly different direction. Thus the light ray will be deflected from its original path. This is the rough picture and now we will proceed with the calculation.

It is convenient to transform to dimensionless coordinates by setting $v = bu$. Then we can write Rq. (7.2.1) in terms of v as:

$$\frac{d^2 v}{d\phi^2} + v = \epsilon\, v^2, \tag{7.2.2}$$

where $\epsilon = 3m/b$. Note that ϵ is dimensionless and for the situation under consideration it is small. For the specific case of a light ray grazing the surface of the Sun, b can be taken to be the radius of the Sun and $m \sim 1.5$ km, the mass of the Sun in length units. Therefore ϵ is few times 10^{-6}, much smaller than unity. Now we are ready to carry out the approximation order by order in the small parameter ϵ. We need only go the first order in ϵ and accordingly we write $v = v_0 + \epsilon\, v_1$. Then writing Eq. (7.2.2) with this form of v we obtain:

$$\frac{d^2 v_0}{d\phi^2} + v_0 + \epsilon\, \frac{d^2 v_1}{d\phi^2} + \epsilon\, v_1 = \epsilon\, (v_0 + \epsilon v_1)^2. \tag{7.2.3}$$

We can now separate out the equations at various orders of ϵ. We write out the equations at order 0 and order 1:

$$\frac{d^2 v_0}{d\phi^2} + v_0 = 0 \tag{7.2.4}$$

$$\frac{d^2 v_1}{d\phi^2} + v_1 = v_0^2. \tag{7.2.5}$$

The solution to the v_0 equation with boundary conditions as chosen is $v_0 = \sin \phi$. This corresponds to $r \sin \phi = b$ which is $y = b$. This is the zero'th order solution to the problem. We now substitute this solution into the second equation and obtain:

$$\frac{d^2 v_1}{d\phi^2} + v_1 = \sin^2 \phi. \tag{7.2.6}$$

The solution to this equation is:

$$v_1 = \frac{2}{3} - \frac{1}{3} \sin^2 \phi, \tag{7.2.7}$$

which then gives on substituting for ϵ, $v = \sin \phi + 2m/b - (m/b) \sin^2 \phi$. We need the asymptotes to this curve which are obtained when $v \to 0$. Dropping the last term as it is of higher order of smallness, we find that the asymptotes are given by

$\sin\phi = -2m/b$ which gives two solutions $\phi = -2m/b,\ \pi + 2m/b$. Thus the deflection angle is $\Delta\phi = 4m/b$. If one puts in the values corresponding to the Sun's mass and radius, we find $\Delta\phi \sim 1.75"$ in agreement with observations.

We remark that using only the equivalence principle for solving this problem gives half this value. This problem has been worked out in Exercise 2.3.2 of Chapter 2.

7.3   The Perihelion Shift in the Orbit of Mercury

In Newton's theory a test particle orbit is a conic section—an ellipse, a hyperbola or a parabola. If the orbit is bound as in the case of planetary motion, then the planet must describe an ellipse (in the absence of other perturbing forces). It was found that the orbit of Mercury is a precessing ellipse. Surely, there are other planets orbiting the Sun which are responsible in part for the precession, but not all of the precession could be explained away. The discrepancy not accounted for is minute—about 43 arc seconds per century. *Although the discrepancy is minute, it cannot be denied.* It was a great triumph for GR that it could exactly explain this discrepancy and GR became a theory to be reckoned with. Figure 7.2 shows a precessing elliptical orbit with the sun (orange disc) at the focus. The orbit is not closed. The precession has been grossly exaggerated for clarity. In this subsection we will establish this result.

We begin with Eqs. (6.4.9) and (6.4.10), then divide one by the other to obtain the following equation:

$$\frac{1}{r^4}\left(\frac{dr}{d\phi}\right)^2 = \frac{E^2 - 1}{L^2} + \frac{2m}{r}\left(\frac{1}{r^2} + \frac{1}{L^2}\right) - \frac{1}{r^2}. \qquad (7.3.1)$$

Changing the variable to $u = 1/r$ as before results in,

$$\left(\frac{du}{d\phi}\right)^2 + u^2 = \frac{E^2 - 1}{L^2} + 2mu\left(u^2 + \frac{1}{L^2}\right). \qquad (7.3.2)$$

This is however somewhat an inconvenient equation. A simpler equation is obtained on differentiating this equation:

$$u'' + u = \frac{m}{L^2} + 3mu^2, \qquad (7.3.3)$$

where in the above equation we have used a short hand notation by denoting differentiation with respect to ϕ with a prime.

Let us now pause and understand the magnitude of the terms in Eq. (7.3.3) for Mercury's orbit. If one ignores the second term in this equation on the RHS, this equation describes just the Newtonian orbit. The quantity L—angular momentum per unit rest mass at infinity—is just the Newtnonian angular momentum L_N/c. This is evident from Eq. (6.4.9) in which the differentiation is with respect to s which

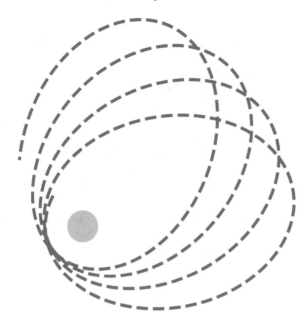

Fig. 7.2 The figure depicts a precessing elliptical orbit with the sun (orange disc) at the focus. The orbit is not closed. The precession has been grossly exaggerated for clarity

is the proper time multiplied by c or $ds = cd\tau$. But since the velocity of Mercury is small compared with c, $ds \approx cdt$. This brings in the factor c or $L_N = cL$. The Newtonian equation is,

$$u'' + u = \frac{m}{L^2} = \frac{GM}{L_N^2} \equiv \frac{1}{l}, \tag{7.3.4}$$

where l is the latus rectum. This equation has the solution:

$$u = \frac{1}{l}\,(1 + e\cos\phi)\,, \tag{7.3.5}$$

Eq. (7.3.5) is the Newtonian solution for the orbit. e is the eccentricity of Mercury's orbit and it is reasonably small so that the orbit can be considered to be roughly circular. Then $l \sim$ radius of the orbit $\sim 6 \times 10^7$ km. Now going back to Eq. (7.3.3) we can see that the second term on the RHS is $\sim 3m/l^2$ because $u \sim 1/l$ while the first term is $1/l$. Thus the second term is smaller by a factor of $3m/l \sim 10^{-7}$ as compared to the first. Recall that the Sun's mass $m \sim 1.5$ km. It is therefore a small perturbation of the orbit and therefore it has an extremely small effect. The essential effect we are interested in is the precession of the elliptical orbit.

Let us therefore write the total solution as $u = u_0 + u_1$, where u_0 is the Newtonian solution given in Eq. (7.3.5). Then to the first order u_1 satisfies the equation:

$$
\begin{aligned}
u_1'' + u_1 &= 3mu_0^2, \\
&= \frac{3m}{l^2}\left(1 + 2e\cos\phi + e^2\cos^2\phi\right).
\end{aligned} \tag{7.3.6}
$$

We remark that the perturbation scheme can be carried out in a systematic way just as we did for the case of deflection of light by the Sun. We could have defined $v = lu$ and written the full equation as $v'' + v = 1 + \epsilon v^2$, where $\epsilon = 3m/l$, then separated out the equation at orders 0 and 1 of ϵ. Carrying out these steps, we would have obtained identical results.

There are now three terms on the RHS of Eq. (7.3.6). The first and third term, that is the constant term and the $\cos^2\phi$ term only produce a bounded oscillatory solution. It is the term in $\cos\phi$ which is responsible for a secular increase in phase and hence for precession of the orbit. Therefore, taking only this term into account, we find that the solution is,

$$
u_1 = \frac{3me}{l^2}\,\phi\sin\phi. \tag{7.3.7}
$$

The above solution can be easily verified by direct substitution. The full solution is,

$$
\begin{aligned}
u &= \frac{1}{l}(1 + e\cos\phi) + \frac{3me}{l^2}\,\phi\sin\phi \\
&= \frac{1}{l}\left[1 + e(\cos\phi + \delta\sin\phi)\right],
\end{aligned} \tag{7.3.8}
$$

where,

$$
\delta = \frac{3m}{l}\,\phi. \tag{7.3.9}
$$

The term in round brackets can be written as $\cos(\phi - \delta)$ where we approximate $\cos\delta \approx 1$ and $\sin\delta \sim \delta$. This term implies precession of the orbit. In one orbit the ϕ goes from 0 to 2π and thus in one orbit the orbit precesses by the amount,

$$
\Delta\phi = \frac{6\pi m}{l} = \frac{6\pi GM}{lc^2}. \tag{7.3.10}
$$

One may now put in the numbers for Mercury and the Sun. More accurately, $l \sim 5.545 \times 10^7$ km and $m \sim 1.48$ km. Also Mercury takes about 88 days to orbit the Sun. Thus in one century Mercury completes about 415 orbits. Multiplying $\Delta\phi$ as given in Eq. (7.3.10) by 415 gives the precession rate of about $43''$ per century in agreement with observations.

7.4 Gravitational Redshift: Pound-Rebka Experiment

Gravitational redshift is a fundamental prediction of GR, which not only provides an important validation of the theory, but is crucial in day-to-day life, specifically in Global Positioning System (GPS) primarily used for navigation. This topic has been covered in the previous chapters and the expressions were derived for Schrawzschild geometry in Chapter 6. Here we describe the first gravitational redshift measurement due to earth's gravitational field.

In 1959 Robert Pound and Glenn Rebka devised an experiment at Harvard University to test the gravitational redshift prediction of General Relativity. They placed atomic gamma ray emitters on top of a tower and the same material at the bottom of the tower which would absorb those photons. However, due to gravitational blue shift, the frequency of the emitted photons increases when they reach the bottom of the tower, which does not match the difference in energy levels required for the transition to occur and hence the photons pass unabsorbed. However, if the emitter atoms are given the right amount of upward velocity (that is, away from the receiver atoms) that could compensate for the gravitational blueshift and then the photons would get absorbed. This is the basic principle for the experiment.

If R is the radius of the earth, the h ($h \ll R$) height of the tower, λ_e the wavelength of the emitted photons and λ_r the wavelength when the photons reach the receiver atoms, then from the gravitational redshift formula, Eq. (6.5.3), we obtain:

$$\frac{\lambda_r}{\lambda_e} = \left[\frac{(g_{00})_r}{(g_{00})_e}\right]^{1/2} = \frac{\left(1 - \frac{2m}{R}\right)^{1/2}}{\left(1 - \frac{2m}{R+h}\right)^{1/2}} \approx 1 + \frac{m}{R+h} - \frac{m}{R} \approx 1 - \frac{mh}{R^2}, \quad (7.4.1)$$

where, as usual $m = GM_\oplus/c^2$, and $M_\oplus$ is the mass of the earth. If the emitter atoms are given a velocity v upwards and if λ_0 is the wavelength of the photons in the rest frame of the atoms, due to special relativistic Doppler shift the effective emitted wavelength λ_e becomes,

$$\lambda_e = \sqrt{\frac{1 + v/c}{1 - v/c}}\lambda_0 \approx (1 + v/c)\lambda_0. \quad (7.4.2)$$

Thus the wavelength of the photons received at the bottom of the tower is,

$$\lambda_r = \left(1 - \frac{mh}{R^2}\right)\left(1 + \frac{v}{c}\right)\lambda_0. \quad (7.4.3)$$

In order for the received photons to have the same wavelength, $\lambda_r = \lambda_0$, so that they could be absorbed by the receiving atoms in their rest frame, requires, to the first order, $v/c = mh/R^2 = GM_\oplus h/(Rc)^2 = gh/c^2$, where g is the acceleration due to gravity at the surface of the earth.

In the set up used by Pound and Rebka, the emitters were kept at a height of $h = 22.5$m with respect to the receiving atoms. Hence, the velocity required for the emitters to satisfy the absorption criteria described above, $v \approx 2.5 \times 10^{-15}c = 7.5 \times 10^{-7}$m/s. However, this velocity is much smaller (by five orders of magnitude) than the recoil velocity of each atom when it emits or absorbs a photon. To alleviate this problem, Pound and Rebka made use of Mössbauer spectroscopy, where the atoms (^{57}Fe in this case) were embedded in a much heavier solid crystal lattice that, as a whole, reacted to the recoil momentum, significantly reducing the recoil velocity. In the experiment, the emitter was mounted on a loudspeaker cone which moved the emitting atoms with sinusoidally varying velocity. When the velocity matched the required value to compensate for the gravitational redshift, the photons would get absorbed by the receiving atoms. This showed a variation in the reading of a scintillation counter kept below the receiver. To rule out spurious correlations, the experiment was performed for different loudspeakers for a range of oscillation frequencies. The results matched the prediction and GR successfully passed yet another test.

7.5   Shapiro Time Delay

An electromagnetic wave such as light or a radio wave experiences an extra time-delay if the path of the light passes close to a gravitating object. This was first pointed out by Shapiro. For example consider say Venus and Earth on opposite sides of the Sun and the null ray from one to the other passes close to the surface of the Sun. Then the null ray takes a little longer to travel the distance as compared to the flat space situation. If a radar beam is sent from Earth to Venus and it reflects of its surface, the total extra time taken by the ray is $\sim 200\mu$sec. We will now carry out the analysis below.

We again use the variational principle given by Eq. (6.4.3) to obtain the photon orbits. The first integrals of the orbits are given in Eq. (6.4.20). Here we need the $\dot{r}$ and $\dot{t}$ equations which we must divide to yield the result:

$$\left(\frac{\dot{r}}{c\dot{t}}\right)^2 \equiv \left(\frac{dr}{cdt}\right)^2 = \left(1 - \frac{2m}{r}\right)^2 \left[1 - \frac{b^2}{r^2}\left(1 - \frac{2m}{r}\right)\right], \qquad (7.5.1)$$

where as before we have put $L/E \equiv b$—the impact parameter which is also a constant of motion. In our situation because the radar ray grazes the surface of the Sun the value of b can be taken to be the radius of the Sun or b is approximately 10^6 km. Since $2m \sim 3$ km we have $b >> 2m$. Further, since the planets are at a distance $\sim 10^8$ km, this distance, say r_1, is much larger than the radius of the Sun and so $r_1 >> b$. We must keep these numbers in mind while performing the integrals so that suitable approximations can be made to simplify the integrals.

The bounce of the ray occurs approximately at $r \approx b$. This is the minimum distance the ray gets near to the Sun. It is convenient to switch to dimensionless variables. Let $x = r/b$, $T = ct/b$ and $\epsilon = 2m/b$. In these variables Eq. (7.5.1) becomes,

$$\left(\frac{dx}{dT}\right)^2 = \left(1 - \frac{\epsilon}{x}\right)^2 \left[1 - \frac{1}{x^2}\left(1 - \frac{\epsilon}{x}\right)\right].\qquad(7.5.2)$$

We now take the square root of this equation and then take its reciprocal. The result is:

$$\frac{dT}{dx} = \left(1 - \frac{\epsilon}{x}\right)^{-1} \left[1 - \frac{1}{x^2}\left(1 - \frac{\epsilon}{x}\right)\right]^{-\frac{1}{2}} \approx \left(1 + \frac{\epsilon}{x}\right)\left(1 - \frac{1}{x^2}\right)^{-\frac{1}{2}},\qquad(7.5.3)$$

where we have used the binomial theorem and kept terms upto the first order in ϵ. We now integrate the above equaion from b to r_1 or in our dimensionless variables from 1 to $x_1 = r_1/b$. We obtain:

$$T = \int_1^{x_1} \frac{x\,dx}{\sqrt{x^2 - 1}} + \epsilon \times \int_1^{x_1} \frac{dx}{\sqrt{x^2 - 1}}$$
$$\equiv T_{\text{Newtonian}} + T_{\text{Shapiro}}.\qquad(7.5.4)$$

The first integral is just $\sqrt{x_1^2 - 1}$ which in terms of r becomes:

$$\sqrt{r_1^2 - b^2} \equiv c \times t_{\text{Newtonian}},\qquad(7.5.5)$$

which is just c times the time taken by the photon to travel from $r = r_1$ to b in flat space, in the absence of gravity. This quantity we identify with the Newtonian value. The second integral gives,

$$T_{\text{Shapiro}} = \epsilon \ln\left(x_1 + \sqrt{x_1^2 - 1}\right).\qquad(7.5.6)$$

This expression gives the time-delay for the photon path from r_1 to b:

$$t_{\text{Shapiro}} = 2m \ln\left(\frac{r_1 + \sqrt{r_1^2 - b^2}}{b}\right) \approx 2m \ln\left(\frac{2r_1}{b}\right).\qquad(7.5.7)$$

We may now compute the full Shapiro delay of the radar pulse travelling from $r_1 \longrightarrow b \longrightarrow r_2$ and back the same way, where r_1 and r_2 are the radial coordinates of the planets. Then using Eq. (7.5.7) from r_1 to b and again from b to r_2 and doubling the result for the return path, we obtain,

$$t_{\text{Shapiro}} \approx 4m \ \ln \ \left(\frac{4r_1 r_2}{b^2} \right) . \tag{7.5.8}$$

Taking the values $r_1 \sim 1.5 \times 10^8$km (Earth-Sun distance), $r_2 \sim 1.08 \times 10^8$km (Venus-Sun distance) and $b \sim 0.7 \times 10^6$km, the radius of the Sun, the above equation results in $t_{\text{Shapiro}} \sim 2.4 \times 10^{-4}$ seconds.

Exercises

1. Global Positioning System (GPS) devices use time delays from different GPS satellites to determine the device's current location. The satellites are equipped with very accurate clocks. Since 1 nano-second time delay corresponds to about 30 cm, the clocks must be very precisely synchronised to provide acceptable accuracy for navigation. It turns out that (special relativistic) time dilation due to the orbital velocity of satellites and (general relativistic) gravitational redshift can lead to tens of microseconds of time delay per day where the ground reference is the earth sea-level. Both special and general relativistic corrections are far greater than the desired precision required for normal navigational purposes. Therefore these effects are important and must be accounted for. This is a good example of how special and general relativity matter in our day to day life. Consider a satellite in a circular orbit of radius r around the earth, $R_\oplus = 6400$ km is the radius of the earth and g_{SL} is the acceleration due to gravity at the sea level.

   (a) Show that the fractional shift in time ΔT_{SR} in the satellite with respect to the GPS device, essentially taken to be at rest on ground, due to the orbital speed of the satellite is:

   $$\frac{\Delta T_{SR}}{T} \approx - \frac{g_{SL} \, R_\oplus^2}{2 \, c^2 \, r} .$$

   [Hint: take the orbital speed of the satellite to be the relative speed between the satellite and the GPS device and assume non-relativistic speeds.]

   (b) Show that the fractional time shift in the satellite with respect to the GPS device (at sea-level) due to redshift caused by earth's gravitational field is

   $$\frac{\Delta T_{GR}}{T} \approx \frac{g_{SL} \, R_\oplus}{c^2} \left(1 - \frac{R_\oplus}{r} \right) .$$

   (c) For a satellite at an altitude of 20, 200 km (i.e., $r = 26, 600$ km) calculate the numerical values of the above time shifts in μs/day units to show that the GR effect is few times stronger than the SR effect, but opposite in sign.

   (d) At what altitude of the satellite does the total time shift ($\Delta T_{SR} + \Delta T_{GR}$) vanish?

Chapter 8
Gravitational Waves

8.1 Introduction

Just as Einstein's equations admit static solutions such as the Schwarzschild, they also admit wave solutions. Thus gravitational wave solutions exist in the theory of general relativity. It is easier to see this in the weak field limit. In the strong field regime there do exist gravitational wave solutions, but we will not consider them here. This is because, astrophysically we expect the waves to be weak, the term "weak" is made more precise later in the text. Gravity is a weak force and from the knowledge we have about astrophysical sources from electromagnetic observations, we expect the waves to be weak. Given their energetics, their distances, etc., the waves can be considered to be in the weak field regime and their propagation, detection etc. can be analysed accordingly. We have already considered the Newtonian approximation which is both weak field and slow motion, but here we need to consider only the weak field approximation and not slow motion—after all the GW, as we will show later, travel at the speed c. In order to make the weak field approximation we need to linearise Einstein's equations. Note that since this is not the Newtonian approximation, because we allow speeds comparable to the speed of light; we retain the time derivatives in the equations.

Also another comment is in order about the index notation. In this chapter we will require the use of two types of indices, (i) spacetime indices which we will denote by Greek letters, α, β, ... etc. and which take values 0, 1, 2, 3 and (ii) purely spatial indices (latin) which we will denote by i, j, k, ... which take values 1, 2, 3. The index 0 corresponds to the time index.

The original version of this chapter was revised: Belated corrections in equations have been updated. The correction to this chapter is available at https://doi.org/10.1007/978-3-030-92335-8_10

8.2 Linearised Gravity

We choose a nearly Cartesian system so that we can write:

$$g_{\alpha\beta} = \eta_{\alpha\beta} + h_{\alpha\beta}\,, \qquad (8.2.1)$$

where $|h_{\alpha\beta}| << 1$ and $\eta_{\alpha\beta}$ is the Minkowski metric tensor with only nonzero diagonal components $(1, -1, -1, -1)$. We can think of $h_{\alpha\beta}$ as a "field" over the Minkowski spacetime described by the metric $\eta_{\alpha\beta}$. This is in the spirit of considering an electromagnetic field described by the electromagnetic vector potential A_α over the Minkowski spacetime. From astrophysical considerations, we typically find $h_{\alpha\beta} \sim 10^{-22}$ or less when the waves reach our detectors. Therefore, we are essentially justified in keeping only the first order terms in $h_{\alpha\beta}$. It is possible that near to the source, $h_{\alpha\beta}$ may not be so small and higher orders are required, but we will not be dealing with such situations in this book.

We start with linearising the Einstein tensor which consists of the Ricci tensor and the scalar curvature which are themselves contractions of the Riemann tensor. Therefore, we first start with the Riemann tensor. Note that,

$$R_{\alpha\mu\beta\nu} = \frac{1}{2}\left(g_{\alpha\nu,\mu\beta} + g_{\mu\beta,\alpha\nu} - g_{\alpha\beta,\mu\nu} - g_{\mu\nu,\alpha\beta}\right) + \Gamma^2 \text{ terms}\,. \qquad (8.2.2)$$

Given the weak field limit we find that $\Gamma \sim o(h)$ and so the product of Gamma terms are of the second and higher order in $h_{\alpha\beta}$ and so can be dropped in this approximation. Also the expression in $g_{\alpha\beta}$ between the parenthesis reduces to the expression with gs replaced by hs because the η_{ik} are constants whose derivatives are zero. Thus in the weak field limit,

$$R_{\alpha\mu\beta\nu} \simeq \frac{1}{2}\left(h_{\alpha\nu,\mu\beta} + h_{\mu\beta,\alpha\nu} - h_{\alpha\beta,\mu\nu} - h_{\mu\nu,\alpha\beta}\right)\,. \qquad (8.2.3)$$

In order to proceed further, that is, compute the Ricci tensor and the scalar curvature, it helps to go over to a new variable,

$$\bar{h}_{\alpha\beta} = h_{\alpha\beta} - \frac{1}{2} h\, \eta_{\alpha\beta}\,, \qquad (8.2.4)$$

where $h = h^\mu_\mu$, is the trace of $h_{\alpha\beta}$. Note that the indices are lowered and raised by the metric tensors $\eta_{\alpha\beta}$ and $\eta^{\alpha\beta}$, because we are keeping terms only to the first order in $h_{\alpha\beta}$. The quantity $\bar{h}_{\alpha\beta}$ is called the trace reverse of $h_{\alpha\beta}$. By making infinitismal coordinate transformations of the form $x'^\mu = x^\mu + \xi^\mu$ where ξ^μ are infinitismal we can choose a coordinate system in which the divergence of $\bar{h}_{\alpha\beta}$ vanishes, that is,

$$\bar{h}^{\alpha\beta}_{,\beta} = 0\,. \qquad (8.2.5)$$

This is called the Lorenz gauge in analogy with the Lorenz gauge in electrodynamics (see Eq. (1.7.17)). It can be shown that such a coordinate system can always we found if the relevant functions are well behaved—continuous, differentiable etc. We show this in the next section. Moreover, just as in electrodynamics, this choice of gauge does not exhaust all freedom, because we can still remain in the gauge by making further transformations where the functions ξ^i satisfy $\Box \xi^i = 0$. So it is really a family of gauges. We will use this freedom in the next subsection to simplify the algebra and thereby in doing so, fix the gauge.

In order to write down Einstein's equations, we further need to calculate the Ricci tensor $R_{\mu\nu}$ and the scalar curvature R in terms of $h_{\alpha\beta}$. The Ricci tensor is given by:

$$\begin{aligned} R_{\mu\nu} &= \eta^{\alpha\beta} R_{\alpha\mu\beta\nu} \\ &= \frac{1}{2} (h^{\beta}{}_{\nu,\mu\beta} + h^{\alpha}{}_{\mu,\alpha\nu} - h_{,\mu\nu} - \eta^{\alpha\beta} h_{\mu\nu,\alpha\beta}) \,. \end{aligned} \qquad (8.2.6)$$

In the Lorenz gauge we have: $\bar{h}^{\mu\nu}{}_{,\nu} = h^{\mu\nu}{}_{,\nu} - \frac{1}{2} h^{,\mu} = 0$,
which gives the relations,

$$h^{\nu}{}_{\mu}{}_{,\nu} = \frac{1}{2} h_{,\mu} \,,$$

$$h^{\beta}{}_{\nu,\mu\beta} = h^{\alpha}{}_{\mu}{}_{,\alpha\nu} = \frac{1}{2} h_{,\mu\nu} \,. \qquad (8.2.7)$$

Recognising that, $\eta^{\alpha\beta} h_{\mu\nu,\alpha\beta}$ is just the d'Alembertian of $h_{\mu\nu}$ that is, $\Box h_{\mu\nu}$ we have,

$$R_{\mu\nu} = -\frac{1}{2} \Box h_{\mu\nu}, \qquad R = \eta^{\mu\nu} R_{\mu\nu} = -\frac{1}{2} \Box h \,. \qquad (8.2.8)$$

The Einstein tensor then becomes,

$$R_{\mu\nu} - \frac{1}{2} \eta_{\mu\nu} R = -\frac{1}{2} \Box \bar{h}_{\mu\nu} \,, \qquad (8.2.9)$$

To arrive at the above results, considerable amount of algebra is required. It is a good excercise for the student to work through the tensor algebra.

Einstein's equations then take the form:

$$\Box \bar{h}_{\mu\nu} = -\frac{16\pi G}{c^4} T_{\mu\nu} \,. \qquad (8.2.10)$$

These are inhomogeneous wave equations with the stress tensor being the source for the wave operator, namely, the $\Box$ on the left hand side.

In latter sections we will consider source free solutions to Eq. (8.2.10) or wave propagation. Later we will consider wave generation where the source term is crucial. While deriving the above results, we had tacitly assumed that it was possible to choose the Lorenz guage. In the next section we show we that this is always possible under ordinary circumstances.

8.3   Choice of the Lorenz Gauge

First we make some general remarks:

- We are free to choose any coordinate system we want in order to solve a physical problem—physics is coordinate independent.
- We will therefore choose that coordinate system which simplifies calculations and extracts the physics in the simplest possible way. We usually do the same in other branches of physics.

Let us start with an arbitrary coordinate system x^μ. Suppose the Lorenz condition Eq. (8.2.5) in the coordinate sytesm x^μ does not hold. We would like to go over to a system of coordinates, say x'^μ, in which the Lorenz gauge condition does hold. We expect the new system to differ only slightly from the old one, since $|h_{\mu\nu}| << 1$. We therefore write:

$$x'^\mu = x^\mu + \xi^\mu, \tag{8.3.1}$$

where ξ^μ is small in the sense $|\xi^\mu_{,\nu}| << 1$. Now our first goal is to find $h'_{\mu\nu}$ in the new coordinate system. For this purpose we require,

$$\frac{\partial x'^\mu}{\partial x^\alpha} = \delta^\mu_\alpha + \xi^\mu_{,\alpha}, \qquad \frac{\partial x^\alpha}{\partial x'^\mu} \simeq \delta^\alpha_\mu - \xi^\alpha_{,\mu}. \tag{8.3.2}$$

In the second of the equations, the inverse has been taken to the first order in $\xi^\mu_{,\nu}$, since this term is small. From these relations, we compute $g'_{\mu\nu}$ in the new coordinate system:

$$
\begin{aligned}
g'_{\mu\nu} &= \frac{\partial x^\alpha}{\partial x'^\mu} \frac{\partial x^\beta}{\partial x'^\nu} g_{\alpha\beta} \\
&= (\delta^\alpha_\mu - \xi^\alpha_{,\mu})(\delta^\beta_\nu - \xi^\beta_{,\nu})(\eta_{\alpha\beta} + h_{\alpha\beta}) \\
&\equiv \eta_{\mu\nu} + h'_{\mu\nu}.
\end{aligned}
\tag{8.3.3}
$$

To the first order in smallness, we obtain:

$$h'_{\mu\nu} = h_{\mu\nu} - \xi_{\mu,\nu} - \xi_{\nu,\mu}. \tag{8.3.4}$$

From this expression we must compute $\overline{h}'_{\mu\nu}$. Thus we have,

$$
\begin{aligned}
h' &= \eta^{\mu\nu} h'_{\mu\nu} = h - 2\,\xi^\mu_{,\mu} \\
\overline{h}'_{\mu\nu} &= \overline{h}_{\mu\nu} - \xi_{\mu,\nu} - \xi_{\nu,\mu} + \eta_{\mu\nu}\,\xi^\alpha_{,\alpha}.
\end{aligned}
\tag{8.3.5}
$$

Taking the divergence we get:

$$\overline{h}'^{\mu\nu}_{,\nu} = \overline{h}^{\mu\nu}_{,\nu} - \xi^{\mu,\nu}_{,\nu} = \overline{h}^{\mu\nu}_{,\nu} - \Box\,\xi^\mu. \tag{8.3.6}$$

Setting $\overline{h}'^{\mu\nu}{}_{,\nu} = 0$ we obtain the following equation for ξ^μ:

$$\Box\, \xi^\mu = \overline{h}^{\mu\nu}{}_{,\nu}\,. \tag{8.3.7}$$

Such an equation in general has a solution if the functions involved are well behaved which is usually the case in the situations we consider. We do not explicitly need the solution for ξ^μ—we just need to know that 4 such functions ξ^μ exist and thereby establishing that a Lorenz gauge can be chosen.

We remark that, this however does not fix the gauge because we can add ζ^μ to ξ^μ where $\Box\zeta^\mu = 0$. In the next section we will obtain plane wave solutions in a source free region and we will further fix the gauge in which the solutions are most simply expressed.

8.4 Plane Wave Solutions

We consider the simplest type of solutions and also the most important type of solutions, namely, the plane wave solutions—the surfaces of constant phase are planes or more precisely, hyperplanes in the background four dimensional Minkowski space. Astrophysically, since the sources are at a very large distance from the detectors, the waves would be plane to a great degree of accuracy.

We set the right hand side in Eq. (8.2.10), namely, the source term to zero, and so we have the equation:

$$\Box\, \overline{h}_{\mu\nu} = 0\,. \tag{8.4.1}$$

We now consider solutions of the form,

$$\overline{h}_{\mu\nu} = A_{\mu\nu}\, e^{-ik_\alpha x^\alpha}\,, \tag{8.4.2}$$

where $A_{\mu\nu}$ is a constant symmetric matrix and k_α is a constant vector called the wave vector and signifies the direction of the wave in the 4 dimensional spacetime. Substituting this solution in the wave equation gives:

$$\Box\, \overline{h}^{\mu\nu} = \eta^{\alpha\beta}\, \overline{h}^{\mu\nu}{}_{,\alpha\beta} = -\eta^{\alpha\beta} k_\alpha k_\beta\, \overline{h}^{\mu\nu} \equiv 0$$
$$\implies \eta^{\alpha\beta} k_\alpha k_\beta = 0\,. \tag{8.4.3}$$

Or $k_\alpha k^\alpha = 0$, which implies that the wave vector k^α is null. Thus, GW waves travel with the speed c which is also the speed of light. This is the first result of GR.

We will further also show that the waves are transverse. The Lorenz gauge condition Eq. (8.2.5), gives,

$$\overline{h}^{\mu\nu}{}_{,\nu} = 0 \implies A^{\mu\nu} k_\nu = 0\,. \tag{8.4.4}$$

Thus GW are transverse.

8.4.1   The Transverse Traceless (TT) Gauge

The above considerations however do not exhaust the gauge degrees of freedom because we can still remain within the Lorenz gauge by further choosing ξ^μ such that $\Box \xi^\mu = 0$ (instead of using ζ^μ we persist with ξ^μ without any cause for confusion). Let us choose $\xi_\mu = B_\mu e^{-i\,k_\alpha x^\alpha}$ then, Eq. (8.3.5) gives,

$$A'_{\mu\nu} = A_{\mu\nu} + i B_\mu k_\nu + i B_\nu k_\mu - i \eta_{\mu\nu} B^\alpha k_\alpha \,. \tag{8.4.5}$$

We see that we automatically remain in the Lorenz guage because,

$$A'_{\mu\nu} k^\nu = A_{\mu\nu} k^\nu + i B_\mu k_\nu k^\nu + i B_\nu k^\nu k_\mu - i \eta_{\mu\nu} k^\nu B^\alpha k_\alpha \equiv 0 \,, \tag{8.4.6}$$

because first two terms on the RHS of Eq. (8.4.6) are zero and the last two terms cancel out. This is no surprise because we are within the Lorenz gauges.

The additional gauge freedom allows us to make $A_{\mu\nu}$ tracefree, that is,

$$A^\mu_{\ \mu} = 0 \,, \tag{8.4.7}$$

and further for any given timelike vector U^μ we are allowed to choose:

$$A_{\mu\nu} U^\nu = 0 \,. \tag{8.4.8}$$

The equations (8.4.7) and (8.4.8) can be easily shown to hold in the new guage by explicitly solving for B_μ in terms of the old $A_{\mu\nu}$. See Exercise 8.1.

Since $A_{\mu\nu}$ can be considered as a symmetric matrix, it has at most 10 independent components. The Lorenz gauge condition $A_{\mu\nu} k^\nu = 0$, Eqs. (8.4.7) and (8.4.8) constitute 8 independent conditions on these 10 components of $A_{\mu\nu}$. Therefore, $A_{\mu\nu}$ has only 2 independent components and these are the physical degrees of freedom of the GW—the two polarisations. The above conditions fix the gauge and is called the transverse traceless gauge or the TT gauge for short. The $A_{\mu\nu}$ matrix is denoted by $A^{TT}_{\mu\nu}$ in this gauge and the corresponding $h_{\mu\nu}$ by $h^{TT}_{\mu\nu}$.

In order to fix ideas, we compute the TT components for the special case of the wave travelling in the z direction. Then, $k^\mu = (k, 0, 0, k)$ and the space time dependence is of the form $e^{ikz-i\omega t}$, where $k = \omega/c$. Physically, U^μ can be chosen as the 4-velocity of the observer/detector. By going to the frame of the observer we have $U^\mu = (1, 0, 0, 0)$ or $U^\mu = \delta^\mu_0$. Then Eq. (8.4.8) implies that $A_{\mu 0} = 0$. Further rotate the frame so that the 3-vector $\mathbf{k}$ points in the z-direction so that $k^\mu = (\omega/c, 0, 0, \omega/c)$, then $A_{\mu\nu} k^\nu = 0 \Longrightarrow A_{\mu z} = 0$. Thus we find that $A_{\mu\nu}$ has possibly, only (x, y) components non-zero. From symmetry $A_{xy} = A_{yx}$ and the tracefree condition gives $A_{yy} = -A_{xx}$. Thus there are only two independent degrees of freedom expressed by A_{xx} and A_{xy}. These are the two independent polarisations of the wave. Thus like electromagnetic waves, gravitational waves are transverse and have two independent polarisations. Putting all these results together we write:

$$A_{\mu\nu}^{TT} = \begin{pmatrix} 0 & 0 & 0 & 0 \\ 0 & A_{xx} & A_{xy} & 0 \\ 0 & A_{xy} & -A_{xx} & 0 \\ 0 & 0 & 0 & 0 \end{pmatrix} \qquad (8.4.9)$$

It is customary to write $A_{xx} = A_+$ and $A_{xy} = A_\times$ for reasons that will become clear later in the text. We therefore write $A_{xx} = A_+$ and $A_{xy} = A_\times$ and the corresponding $h_+ = A_+ e^{-ik_\alpha x^\alpha}$ and $h_\times = A_\times e^{-ik_\alpha x^\alpha}$. We can associate polarisation tensors corresponding to each polarisation (just as there are polarisation vectors for the electromagnetic wave):

$$e_{\mu\nu}^+ = \begin{pmatrix} 0 & 0 & 0 & 0 \\ 0 & 1 & 0 & 0 \\ 0 & 0 & -1 & 0 \\ 0 & 0 & 0 & 0 \end{pmatrix}, \qquad e_{\mu\nu}^\times = \begin{pmatrix} 0 & 0 & 0 & 0 \\ 0 & 0 & 1 & 0 \\ 0 & 1 & 0 & 0 \\ 0 & 0 & 0 & 0 \end{pmatrix}. \qquad (8.4.10)$$

These are the plus ($+$) and cross ($\times$) polarisation tensors. Each of the polarisation tensors describe linear polarisation again denoted by $+$ and $\times$. The nomenclature will become clear in the next section when we consider the effect of each polarisation separately on a ring of test particles. The general wave is a linear combination of the two polarisations and can be written in terms of the polarisation tensors:

$$h_{\mu\nu}^{TT} = h_+ e_{\mu\nu}^+ + h_\times e_{\mu\nu}^\times. \qquad (8.4.11)$$

Here we have polarisation *tensors* because we have a tensor wave. The polarisations can have a phase difference between them which is realised by the fact that the A_+ and $A_\times$ can be complex quantities and so the phases are be accommodated in the phases of the complex numbers.

The polarisation tensors are orthogonal in the sense that their scalar product is zero:

$$e_{\mu\nu}^+ e^{\times \mu\nu} = 0. \qquad (8.4.12)$$

In the above discussion we have restricted ourselves to a specific frequency ω of the wave. These results can be easily generalised to a wave coming from a fixed direction, as a general function of t. The result follows on taking the inverse Fourier transform with respect to ω. Then the GW amplitudes $h_{+,\times}$ become functions of t, that is, now we have $h_+(t)$, $h_\times(t)$, $h_{\mu\nu}^{TT}(t)$ etc. Much of the above discussion follows (Schutz 1995). For further reading we refer the reader to Misner et al. [1973] and Schutz [1995].

8.4.2   TT—Gauge as an Algebraic Projection

In practise it may seem difficult to transform to the TT gauge because one would have to find four functions $\xi^\mu(x^\alpha)$ and then use these to find $h^{TT}_{\mu\nu}$. But we show below that this can be conveniently achieved by simply projecting the tensor $h_{\mu\nu}$, which may not be in the TT gauge, with a projection operator. That is, the TT gauge computation can be performed algebraically.

Consider a wave travelling in a general spatial direction n^i, where n^i is a unit spatial vector $n_i n^i = -1$. Let us take $k^\alpha = (\omega, \omega n^i)$. We set $\omega = 1$ for convenience (this could be done by choosing the unit of time appropriately—choose the period of the wave divided by 2π as the time unit), because in any case it does not affect the end result. Thus we write $k^\alpha = (1, \ n^i)$ and $k_\alpha = (1, \ n_i)$ where we have $n_i = -\delta_{ij}n^j$. We note that the metric on the spatial sections is $-\delta_{ij}$.

Select a coordinate system in which the Lorentz gauge condition is valid. Then $A_{\alpha\beta}k^\beta = 0$. This gives the equations:

$$A_{00} + A_{0j}n^j = 0, \quad A_{i0} + A_{ij}n^j = 0. \tag{8.4.13}$$

Both the above equations imply:

$$A_{00} = -A_{0i}n^i = A_{ij}n^i n^j. \tag{8.4.14}$$

Also we have:

$$\mathrm{Tr}(A) = A^\alpha_{\ \alpha} = A_{00} - \delta^{ij} A_{ij} = -(\delta^{ij} - n^i n^j) A_{ij} \equiv P^{ij} A_{ij}, \tag{8.4.15}$$

where we have defined a projection operator:

$$P^i_{\ j} = \delta^i_j + n^i n_j. \tag{8.4.16}$$

It is easy to check that the projection operator satisfies:

$$P^i_{\ j} P^j_{\ k} = P^i_{\ k}, \qquad P^i_{\ j} n^j = 0. \tag{8.4.17}$$

Note that P^i_j projects vectors orthogonal to n^i in 3-space. Using the metric $-\delta_{ij}$ we obtain the contravariant and covariant forms of the projection operator. We then obtain,

$$P^{ij} = -(\delta^{ij} - n^i n^j), \qquad P_{ij} = -(\delta_{ij} - n_i n_j). \tag{8.4.18}$$

From $A'^\alpha_{\ \alpha} = 0$ and Eq. (8.4.5) we obtain, $\mathrm{Tr}(A) - 2i\,B_\alpha k^\alpha = 0$. Thus,

$$i B_\alpha k^\alpha = \frac{1}{2}\mathrm{Tr}(A) = \frac{1}{2}P^{ij} A_{ij}. \tag{8.4.19}$$

From $A'_{\alpha\beta}\, U^\beta = 0$ and Eq. (8.4.5) and further choosing $U^\alpha = (1, 0, 0, 0)$ we obtain:

$$A'_{\alpha 0} = A_{\alpha 0} + i\, B_\alpha k_0 + i\, B_0 k_\alpha - i\, (B_\mu k^\mu)\, U_\alpha = 0\,. \qquad (8.4.20)$$

Separating the time and space components of B_α and using Eq. (8.4.19), we get,

$$i\, B_0 = -\frac{1}{2}\, A_{00} + \frac{1}{4}\, P^{lm}\, A_{lm} = -\frac{1}{2}\, A_{lm} n^l n^m + \frac{1}{4}\, P^{lm}\, A_{lm}\,,$$

$$i\, B_j = A_{jk} n^k + \frac{1}{2} A_{lm}\, n^l n^m n_j - \frac{1}{4}\, P^{lm}\, A_{lm}\, n_j\,. \qquad (8.4.21)$$

Since as seen before, $A'_{\alpha\beta}$ is only spatial, we need to only consider A'_{ij}. Thus,

$$A'_{ij} = A_{ij} + i\, B_i n_j + i\, B_j n_i + i\, \delta_{ij}\, B_\alpha k^\alpha\,. \qquad (8.4.22)$$

From Eq. (8.4.21) we can substitute for B_j in the above equation. The result is:

$$A'_{ij} = A_{ij} + [A_{ik} n^k + \frac{1}{2} n^k n^l n_i - \frac{1}{4} P^{lm} A_{lm} n_i] n_j$$

$$+ [A_{jk} n^k + \frac{1}{2} n^k n^l n_j - \frac{1}{4} P^{lm} A_{lm} n_j] n_i$$

$$+ \frac{1}{2}\, \delta_{ij}\, P^{lm}\, A_{lm}\,. \qquad (8.4.23)$$

Simplifying the above equation leads to:

$$A'_{ij} = A_{ij} + A_{im} n^m n_j + A_{jl} n^l n_i + A_{lm} n^l n^m n_i n_j - \frac{1}{2} n_i n_j\, P^{lm}\, A_{lm} + \frac{1}{2}\, \delta_{ij}\, P^{lm}\, A_{lm}$$

$$= \delta_i^l \delta_j^m\, A_{lm} - \delta_i^l n^m n_j\, A_{lm} + \delta_j^m n^l n_i\, A_{lm} + n_i n_j n^l n^m\, A_{lm} - \frac{1}{2} P_{ij}\, P^{lm}\, A_{lm}$$

$$= (\delta_i^l + n^l n_i)(\delta_j^m + n^m n_j) A_{lm} - \frac{1}{2} P_{ij}\, P^{lm}\, A_{lm}$$

$$= P_i^l P_j^m\, A_{lm} - \frac{1}{2} P_{ij}\, P^{lm}\, A_{lm}$$

$$\equiv \left[P_i^l P_j^m - \frac{1}{2} P_{ij}\, P^{lm} \right] A_{lm}\,. \qquad (8.4.24)$$

This establishes the required result.

However, the projection operator P_j^i as defined above has limitations. In order to apply this operator to a superposition of waves, the wave vectors must be parallel. So we can use this operator on h_{ij} in the time domain by taking inverse Fourier transform only in frequency as long as the wave has a fixed spatial direction. This is usually the case for isolated sources. Secondly, difficulties are encountered in the non-vacuum case, when the propagation takes place in a region which is not source free. Then the operator needs modification. See Racz [2009] and Ashtekar and Bonga [2017] for detailed discussions. In this book we will only encounter situations when the above Eq. (8.4.24) is valid.

8.5 Quadrupole Formula

We are now ready with the mathematical tools necessary to address the fundamental question, how gravitational waves are generated by asymmetrically moving masses. To do so, we will need to solve Eq. (8.2.10) retaining the source term on the right. Though this tensor equation may seem intimidating, the solution is remarkably similar to the standard retarded potential expression one encounters in electrodynamics. However some more involved calculations are necessary to bring the formula to an easy to use form (Flanagan and Hughes 2005).

The solution to Eq. (8.2.10), $\Box \bar{h}_{\mu\nu} = -(16\pi G/c^4)T_{\mu\nu}$, can be obtained using the radiative Green's functions for inhomogeneous wave equation described in many standard textbooks on electrodynamics (Jackson 1998). Let the observer's postition be denoted by $\mathbf{x}$ while source position by $\mathbf{x}'$. Then $|\mathbf{x} - \mathbf{x}'|$ is the distance between the source and the observer. Then the solution reads,

$$\bar{h}_{\mu\nu}(t, \mathbf{x}) = -\frac{4G}{c^4} \int d^3\mathbf{x}' \frac{[T_{\mu\nu}(t', \mathbf{x}')]_{\text{ret}}}{|\mathbf{x} - \mathbf{x}'|}, \tag{8.5.1}$$

where $[T_{\mu\nu}(t', \mathbf{x}')]_{\text{ret}}$ is the retarded energy-momentum tensor evaluated at an earlier epoch t', where $t' = t - |\mathbf{x} - \mathbf{x}'|/c$.

We first simplify this formula by assuming that the size of the source is much smaller than its distance from the observer. If the source lies in a region near the origin, this implies, $|\mathbf{x}'| \ll |\mathbf{x}| =: r$. Then one an replace $|\mathbf{x} - \mathbf{x}'|$ by $|\mathbf{x}|$ in the denominator of Eq. (8.5.1), which introduces a fractional error of the order $|\mathbf{x} - \mathbf{x}'|/|\mathbf{x}| \sim L/r$, where L is the size of the source. This is generally a very good approximation. For the typical case of the stellar mass binary black holes detected so far by the gravitational wave detectors, the maximum distance between the binary components (L) is of the order of a thousand kilometers when the signal enters the frequency band of the detector, while the distances (r) to these binaries from the earth are typically millions of light years. Thus $L/r \lesssim 10^{-15}$.

Replacing $|\mathbf{x} - \mathbf{x}'|$ by $|\mathbf{x}|$ in the argument of $[T_{\mu\nu}(t', \mathbf{x}')]_{\text{ret}}$ in the numerator of Eq. (8.5.1), however, requires more thought. This approximation leads to an error in the retarded time, $t' = t - |\mathbf{x} - \mathbf{x}'|/c$, of the order L/c. If a significant component of the source is moving rapidly with velocity v, the timescale for a non-negligible change to happen in the source configuration is L/v. Therefore, one must have $L/c \ll L/v$, that is, $v \ll c$ in order for this approximation not to significantly affect the phase of the estimated signal. Thus this simplification is valid only in the "slow-motion" approximation. In practice, the compact objects can attain velocities as high as $c/2$ before merger, hence a more careful computation must be performed to estimate the waveform in this regime. However, since most of the cycles of the binaries occur with lower velocities, even after entering the detector's sensitive frequency band, the last few cycles contain relatively less power. So on the whole, this approximation is very useful in providing a reasonably accurate waveform for the signal, even for a wide range of masses. However, this approximation may not work so well for large stellar mass black holes which all together have a handful of cycles in the detector band just around the merger.

With the foregoing approximations, Eq. (8.5.1) simplifies to,

$$\overline{h}_{\mu\nu}(t, \mathbf{x}) = -\frac{4G}{c^4 r} \int d^3\mathbf{x}' \, T_{\mu\nu}(t - r/c, \mathbf{x}') . \tag{8.5.2}$$

This expression can be further simplified utilizing the divergence-free and symmetric properties of the energy momentum tensor. We now use these properties to further simplify the formula.

One notes that,

$$\partial_0 T^{00} + \partial_i T^{0i} = 0 , \tag{8.5.3}$$

$$\partial_0 T^{0i} + \partial_j T^{ij} = 0 . \tag{8.5.4}$$

Thus, taking derivative of the first equation with respect to $x^0 (= ct)$ and the second one with respect to x^i and combining terms, one gets,

$$\partial_0^2 T^{00} = -\partial_0 \partial_i T^{0i} = \partial_i \partial_j T^{ij} . \tag{8.5.5}$$

Then multiplying both sides of the equation with $x^k x^l$, the left hand side becomes,

$$x^k x^l \partial_0^2 T^{00} = \partial_0^2 (x^k x^l T^{00}) . \tag{8.5.6}$$

The right hand side of the equation, $x^k x^l \partial_i \partial_j T^{ij}$, requires some more massaging to bring it to a desirable form. One observes that,

$$\partial_i \partial_j (x^k x^l T^{ij}) = x^k x^l \partial_i \partial_j T^{ij} + 2(x^k \partial_i T^{il} + x^l \partial_i T^{ik}) + 2T^{kl} \tag{8.5.7}$$

$$= x^k x^l \partial_i \partial_j T^{ij} + 2\partial_i (x^k T^{il} + x^l T^{ik}) - 2T^{kl} . \tag{8.5.8}$$

Hence, one can write,

$$T^{kl} = \frac{1}{2} \partial_0^2 (x^k x^l T^{00}) + \partial_i (x^k T^{il} + x^l T^{ik}) - \partial_i \partial_j (x^k x^l T^{ij}) . \tag{8.5.9}$$

Substituting this in Eq. (8.5.2) and raising the indices as required, we write,

$$\overline{h}^{kl}(t, \mathbf{x}) = -\frac{4G}{c^4 r} \int d^3\mathbf{x}' \left[\frac{1}{2} \partial_0^2 (x'^k x'^l T^{00}) + \partial_i (x'^k T^{il} + x'^l T^{ik}) - \partial_i \partial_j (x'^k x'^l T^{ij}) \right]_{\text{ret}} . \tag{8.5.10}$$

The last two terms in the volume integral are divergence terms, that is, terms with ∂_i operating on functions. The integrals involving such terms can be converted into surface integrals which can be made zero if the sources are bounded, that is, if the sources can be surrounded by a closed surface which lies entirely in vacuum. If the sources lie in a bounded region, we can choose the surface sufficiently large to satisfy this condition so that the T^{kl} vanishes at the bounding surface. Setting these terms to zero, we arrive at:

$$\overline{h}^{kl}(t, \mathbf{x}) \;=\; -\frac{2G}{c^4 r}\partial_0^2 \int d^3\mathbf{x}'\, x'^k x'^l T^{00}(t - r/c, \mathbf{x}'). \tag{8.5.11}$$

Using $T^{00} = \rho c^2$, the density of mass-energy, and introducing the second moment of mass-energy,

$$I^{kl}(t') \;=\; \int d^3\mathbf{x}'\, x'^k x'^l \rho(t', \mathbf{x}'), \tag{8.5.12}$$

we write,

$$\overline{h}_{kl}(t, \mathbf{x}) \;\simeq\; -\frac{2G}{c^4 r}\, \ddot{I}_{kl}(t - r/c), \tag{8.5.13}$$

where the dot represents the time derivative. Note that a factor of c^2 has cancelled out since $\partial_0^2 = (1/c^2)\partial/\partial t^2$.

The final step is to project the quantities in Eq. (8.5.13) to the TT gauge. This makes the quantities trace-free, in which case, $\overline{h}^{kl} = h^{kl}$. Thus, for plane waves travelling in the spatial direction n^i we use the projection operator P_j^i defined in Eq. (8.4.16). Another quantity which we use later is reduced quadrupole moment tensor,

$$\mathcal{I}^{kl} \;=\; I^{kl} - \frac{1}{3}\delta^{kl} I, \tag{8.5.14}$$

where $I = I_k^k$ is the trace of I^{kl}. We thus arrive at the the famous quadrupole formula,

$$h_{kl}^{TT}(t, \mathbf{x}) \;=\; -\frac{2G}{c^4 r}\, \ddot{\mathcal{I}}_{kl}^{TT}(t - r/c), \tag{8.5.15}$$

where the superscript TT on $\mathcal{I}_{kl}$ denotes its tranverse-traceless projection. Certain remarks about the formula are in order. First, the gravitational wave amplitude falls with distance as $1/r$, just like with propagating electromagnetic fields. The energy flux therefore falls off as $1/r^2$. Although, the rate of change of quadrupole moment can be very large for compact binary stars just before merger, the G/c^4 factor ($\sim 8.3 \times 10^{-44}$ in SI units) makes the strain very small, thus making it very difficult to detect the waves. It is also clear that the leading order emission of gravitational waves requires a non-zero second order time derivative of the mass-energy quadrupole moment. Hence, if an object is axially symmetric, any spin about the axis of symmetry will not lead to any GW emission. Thus a spinning sphere with uniform density distribution will not emit GW. Similarly, a spherically symmetric collapse of a star will not produce any GW.

8.6 Energy Carried by Gravitational Waves

Like electromagnetic waves, gravitational waves also carry energy. This can be understood by the famous thought experiment, the sticky bead argument, put forward by the legendary physicist Richard Feynman. Consider two beads that are separated and can slide on a rigid rod. If gravitational waves pass through the system, the beads, being loosely coupled to the rod, will slide over the rod, since the rod can only deform insignificantly due to its internal strong elastic forces. Therefore as the beads slide, frictional forces between the beads and the rod create heat. Since gravitational waves are the only source of energy in this system, this thought experiment shows that gravitational waves carry energy and that they can transfer their energy to matter.

Defining energy of gravitational waves, however, requires understanding the subtleties involved in general relativity. Since the equivalence principle allows gravitation to disappear in infinitesimally small space-time regions, it may not be possible to define energy locally. However, Isaacson [1968] proposed a prescription to arrive at a definition of energy carried by GW, by averaging over a finite volume in space-time. Like other forms of energy, GW energy density also affects the metric. This fact can be utilised to define an effective stress-energy tensor for GW from Einstein equations. This procedure requires expansion of the Ricci tensor to the second order in the metric perturbations. This argument is valid in the limit of the wavelength much shorter than the curvature scale of the background; that is, for rapidly oscillating waves, where the oscillatory first order perturbations average out to zero. These considerations lead to the following effective stress-energy tensor for GW (Hartle 2003),

$$ t_{\mu\nu}^{\text{GW}} = -\frac{c^4}{8\pi G} \left[\langle R_{\mu\nu}^{(2)} \rangle - \frac{1}{2} g_{\mu\nu}^{(0)} \langle R^{(2)} \rangle \right], \tag{8.6.1} $$

where, $g_{\mu\nu}^{(0)}$ is the background metric and $R_{\mu\nu}^{(2)}$ and $R^{(2)}$ are respectively parts of the Ricci tensor and Ricci scalar which are second order in metric perturbations $h_{\mu\nu}$. In the TT gauge, this formula simplifies to (Misner et al. [1973]),

$$ t_{\mu\nu}^{\text{GW}} = \frac{c^4}{32\pi G} \langle \partial_\mu h^{\text{TT}ij} \, \partial_\nu h_{ij}^{\text{TT}} \rangle. \tag{8.6.2} $$

The above formula allows us to estimate the energy flux of a plane polarised monochromatic gravitational wave in the direction of propagation in TT gauge (Hartle 2003). If the coordinates are so chosen that the direction of propagation is along the z-axis and the wave is '+' polarised, we write, [Sect. 8.4.1],

$$ h_{xx}^{\text{TT}} = -h_{yy}^{\text{TT}} = h \cos(\omega t - kz), \tag{8.6.3} $$

where $\omega = ck$ is the angular frequency of the wave. As $cT_{03} \equiv T_{tz}$ represents energy flux along the direction z [see Sec. 5.2], considering the GW part of the stress-energy tensor, one gets (with $\partial_0 = \partial_t/c$, the factor of c cancels out),

$$T_{tz} = \frac{c^4}{32\pi G} \left\langle \frac{1}{c} \partial_t h_{\mathrm{TT}}^{ij} \partial_z h_{ij}^{\mathrm{TT}} \right\rangle = \frac{c^4}{32\pi G} 2h^2 \frac{\omega k}{c} \left\langle \cos^2(\omega t - kz) \right\rangle . \qquad (8.6.4)$$

Since the average of $\cos^2$ over several cycles is $1/2$, the energy flux in this case takes the simple form,

$$\mathcal{F}_{\mathrm{GW}} = cT_{tz} = \frac{c^3 \omega^2 h^2}{32\pi G} . \qquad (8.6.5)$$

Along the same lines, substituting the quadrupole formula, Eq. (8.5.15), in Eq. (8.6.2) and integrating over all directions, the luminosity of a source in the TT gauge can be expressed as Hartle [2003],

$$L_{\mathrm{GW}} = \frac{1}{5} \frac{G}{c^5} \left\langle \dddot{I}_{ij} \dddot{I}^{ij} \right\rangle . \qquad (8.6.6)$$

For a compact binary with equal component masses $m_1 = m_2 = M/2$, in the Newtonian limit, the above formula reduces to,

$$L_{\mathrm{GW}} = \frac{2}{5} \frac{G}{c^5} M^2 a^4 \Omega^6 , \qquad (8.6.7)$$

where a is the separation distance between the black holes and Ω is the orbital angular frequency. For such an equal mass binary, we know that $GM/(2a^2) = \Omega^2(a/2)$. For black holes, if $a = nm$, where $m = GM/c^2$ is the Schwarschild radius of each black hole of mass $M/2$, one gets, $\Omega^2 = c^2/n^3 m^2$. Thus the frequency of the gravitational waves, which is twice the orbital frequency, becomes,

$$f = \frac{\omega}{2\pi} = \frac{\Omega}{\pi} = \frac{c}{\pi n^{3/2} m} . \qquad (8.6.8)$$

The first detected of binary black hole merger, GW150914 (LIGO Scientific Collaboration et al. 2016), with component masses 30 and 35 solar masses, has the maximum luminosity of $\sim 200 M_\odot c^2/s \approx 3.6 \times 10^{56}$ erg/s in the frequency range $\sim 100 - 200$ Hz. We can try to understand these numbers using the formulae derived above. Since the luminosity peaks in the last few cycles, we can get a rough estimate of the maximum emission frequency by taking $n \sim 3$ (that is, the black holes are three Schwarzschild radius away). Further, setting $M \sim 65 M_\odot$, which corresponds to $m \approx 100$ km, the approximate GW frequency becomes $f \sim 180$ Hz, which is consistent with the observations. To estimate the peak luminosity in this case, substituting the above Ω in Eq. (8.6.7), one gets,

$$L_{GW} = \frac{2}{5} \left(\frac{m}{a}\right)^5 \frac{c\,(M/M_\odot)}{m} M_\odot c^2 = \frac{2}{5} \frac{c\,(M/M_\odot)}{n^5 m} M_\odot c^2 = \frac{2}{5} \frac{Mc^3}{n^5 m}. \quad (8.6.9)$$

For GW150914, using the above approximations, one finds $L_{GW} \sim 330 M_\odot c^2$, which is in the same ballpark, though both the frequency and luminosity are on the higher side due to our specific choice of distance between the black holes. If the distance is increased, by setting say $n = 3.5$, a closer match with observation can be found. The formula could also be expressed in an alternative form for numerical calculations in CGS units (Hartle [2003]),

$$L_{GW} \approx 1.9 \times 10^{33} \left(\frac{M}{M_\odot} \frac{1\,\text{hour}}{P}\right)^{10/3} \text{erg/s}. \quad (8.6.10)$$

Since GW frequency for equal mass binaries is twice the orbital frequency, the orbital period corresponds to 10–20 ms. Thus the above formula yields an energy release of $\sim 30 - 300\ M_\odot c^2/\text{s}$, which is consistent with the observed result. Note that solar luminosity is $L_\odot \approx 3.8 \times 10^{33}$ erg/s, which is 23 orders of magnitude less! However, since GW150914 was about 400 Mpc $\sim 8 \times 10^{13}$ AU away, the flux is reduced by a factor of $\sim 6.4 \times 10^{27}$. Thus for such an event, the amount of energy received at the earth is relatively tiny. This is also the case typically, for energetic transient events in the universe.

8.7 GWs from an Inspiraling Binary in Circular Orbit

We will consider the stars as mass points with masses m_1 and m_2 and further assume the masses of the binaries to be equal, that is, $m_1 = m_2 \equiv m$. In Exercise 8.3 we will indicate how to generalise to the case of unequal masses.

We will proceed step by step. We will first begin with binaries in a circular orbit, orbiting with constant angular frequency.

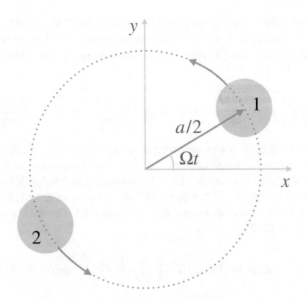

Fig. 8.1 Schematic diagram of an equal mass binary system in circular orbit. The distance between the two binary stars is a and the orbital phase is Ωt

8.7.1 GW Amplitudes for a Equal Mass Binary in Circular Orbit with Constant Angular Frequency

Let the motion of the stars considered as point masses each of mass $M/2$ be described in $(x - y)$ plane with their centre of mass at the origin, aas depicted in Fig. 8.1. We express the masses in terms of the total mass M, because the total mass M appears in Kepler's third law and GW amplitudes are ultimately expressible easily in terms of M. Let the distance between the stars be given by a so that each mass point is at a distance $a/2$ from the origin—each star describes a circular orbit with radius $a/2$ and centre as the origin and with orbital angular velocity Ω. The GW frequency will be denoted by ω. We will first derive the waveform amplitudes $h_+, h_\times$. For this purpose, we will use Eq. (8.5.15). In the next subsection, we will adiabatially evolve the orbit as the system loses energy through GW emission.

Derivation 18 *The coordinates of the point masses are given by:*

$$x_1 = \frac{1}{2} a \cos \Omega t = -x_2, \quad y_1 = \frac{1}{2} a \sin \Omega t = -y_2. \tag{8.7.1}$$

The non-zero quadrupole tensor components are given by:

$$I_{xx} = m_1 x_1^2 + m_2 x_2^2 = \frac{1}{4} M a^2 \cos^2 \Omega t = \frac{1}{8} M a^2 (1 + \cos 2\Omega t),$$

$$I_{yy} = m_1 y_1^2 + m_2 y_2^2 = \frac{1}{4} M a^2 \sin^2 \Omega t = \frac{1}{8} M a^2 (1 - \cos 2\Omega t),$$

$$I_{xy} = m_1 x_1 y_1 + m_2 x_2 y_2 = \frac{1}{4} M a^2 \cos \Omega t \sin \Omega t = \frac{1}{8} M a^2 \sin 2\Omega t. \tag{8.7.2}$$

We now need to take the second time derivative of I_{ij}. Accordingly writing $\ddot{I}_{ij}$ as a 3×3 matrix we have:

$$\ddot{I} = -\frac{1}{2} M a^2 \Omega^2 \begin{bmatrix} \cos 2\Omega t & \sin 2\Omega t & 0 \\ \sin 2\Omega t & -\cos 2\Omega t & 0 \\ 0 & 0 & 0 \end{bmatrix}. \tag{8.7.3}$$

The h_{ij} is the just $2G/c^4 r$ times $\ddot{I}_{ij}$, where r is the distance to the binary system from the observer. For an observer on the z-axis the h_{ij} is already in the TT form and hence for this observer, the GW amplitudes are given by:

$$h_+ = +\frac{GM}{c^4 r} a^2 \Omega^2 \cos 2\Omega t,$$

$$h_\times = +\frac{GM}{c^4 r} a^2 \Omega^2 \sin 2\Omega t. \tag{8.7.4}$$

We can alternatively use the Kepler's third law $a^3 \Omega^2 = GM$ and write the amplitudes in terms of the separation a:

$$h_+ = + \left(\frac{GM}{c^2 r} \right) \left(\frac{GM}{c^2 a} \right) \cos 2\Omega t,$$

$$h_\times = + \left(\frac{GM}{c^2 r} \right) \left(\frac{GM}{c^2 a} \right) \sin 2\Omega t, \tag{8.7.5}$$

where the amplitudes are written in terms of the ratios of M in length units and distances r and a.

Next we compute the amplitudes for an off-axis observer. Let the observer be located in the direction at an angle ι with the z-axis. We only indicate main steps in this calculation. The first step is to apply a rotation matrix $R(\iota)$ to the I tensor.

Derivation 19 *Wave amplitudes for an off-axis observer:*

$$R(\iota) = \begin{bmatrix} 1 & 0 & 0 \\ 0 & \cos\iota & \sin\iota \\ 0 & -\sin\iota & \cos\iota \end{bmatrix}. \tag{8.7.6}$$

The rotation produces a matrix $I' = R(\iota)\, I\, R^T(\iota)$ which we have not explicitly written here. Now take the transverse part by setting the third column and row of I' to zero. Then subtract out the trace to make it traceless and so obtain $\ddot{I}'^{TT}$ (see Exercise 2). From this the GW amplitudes are easily obtained as:

$$h_+ = + \left(\frac{GM}{c^2\,r}\right)\left(\frac{GM}{c^2\,a}\right)\left(\frac{1+\cos^2\iota}{2}\right)\cos 2\Omega t\,,$$

$$h_\times = + \left(\frac{GM}{c^2\,r}\right)\left(\frac{GM}{c^2\,a}\right)\cos\iota\,\sin 2\Omega t\,. \tag{8.7.7}$$

8.7.2   GWs from an Inspiraling Binary

We have obtained the GW amplitudes for the case where the stars remain at a constant distance from each other and revolve around each other with constant angular frequency Ω. However, this is not the case; the stars do not remain at a constant distance from each other, because they lose energy due to the emission of GW according to the formula given in Eq. (8.6.7). If the effect on the orbit is small in the sense that the radius of the orbit changes slowly as compared to the time-scale given by the orbital period $2\pi/\Omega$, that is $\delta a = \dot{a} \cdot 2\pi/\Omega << a$, then we can consider the orbital frequency Ω to be nearly constant for each orbit. The frequency Ω also does not remain constant but changes slowly or adiabatically. We therefore use Eq. (8.6.7) to compute the energy loss for an orbit or few orbits.

Derivation 20 *We begin with Eq. (8.6.7):*

$$L_{GW} = \frac{2}{5} \frac{GM^2}{c^5} a^4 \Omega^6 \equiv -\frac{dE}{dt},$$

(8.7.8)

where E is the total (Newtonian) energy of the binary system. We will now obtain the time rate of change of a. For this purpose we need to write E in terms of a.

$$E = T + V = 2 \times \left(\frac{1}{16} M a^2 \Omega^2 \right) - \frac{GM^2}{4a}$$

$$= -\frac{1}{8} \frac{GM^2}{a},$$

(8.7.9)

where we have made use of the Kepler's third law $a^3 \Omega^2 = GM$. Differentiating the above relation for E with respect to a and equating it to the rate of energy loss, furnishes a differential equation for a:

$$a^3 \frac{da}{dt} = -\frac{16}{5} \frac{G^3 M^3}{c^5},$$

(8.7.10)

where we have again used the Kepler's third law to eliminate Ω. Integrating this equation immediately obtains for us the coalescence time t_c. If the initial separation is a_0 at time $t = 0$, then a becomes zero when the stars coalesce at $t = t_c$. Thus,

$$t_c = \frac{5}{64} \frac{c^5}{G^3 M^3} a_0^4 = \frac{5}{64} \frac{GM}{c^3} \left(\frac{a_0}{GM/c^2} \right)^4 = \frac{5}{64} \frac{GM}{c^3} \left(\frac{\pi GM f_0}{c^3} \right)^{-8/3}.$$

(8.7.11)

Here f_0 is the GW frequency at $t = 0$. In general, the GW frequency is given by $f = \Omega/\pi$.

We now obtain the evolution of the separation $a(t)$ and $\Omega(t)$ and hence the waveform. We integrate Eq. (8.7.10) from t to t_c and a from $a(t)$ to 0. Also from Kepler's law we can immediately obtain $\Omega(t)$ from the expression for $a(t)$. Further integration of $\Omega(t)$ yields the phase $\phi(t)$. The results are:

$$a(t) = \frac{GM}{c^2} \left[\frac{64}{5} \frac{c^3(t_c - t)}{GM} \right]^{1/4},$$

$$\Omega(t) = \frac{c^3}{GM} \left[\frac{64}{5} \frac{c^3(t_c - t)}{GM} \right]^{-3/8},$$

$$\phi(t) = \phi_c - \int_t^{t_c} \Omega(t) \, dt = \phi_c - \left[\frac{2^{6/5} c^3(t_c - t)}{5GM} \right]^{5/8},$$

(8.7.12)

where ϕ_c is the coalescence phase. (The ugly factor $2^{6/5}$ disappears from the expession for the phase, if we replace the total mass M by the chirp mass $\mathcal{M}$. For equal masses $M = 2^{6/5} \mathcal{M}$. See Exercise 3.)

Collecting everything together the GW amplitudes are given by,

$$h_+(t) = + \mathcal{A}(t) \, \frac{1 + \cos^2 \iota}{2} \, \cos 2 \, \phi(t) \,,$$
$$h_\times(t) = + \mathcal{A}(t) \, \cos \iota \, \sin 2 \, \phi(t) \,, \qquad (8.7.13)$$

where now the evolving phase $\phi(t)$ is given by Eq. (8.7.12). The evolving amplitude $\mathcal{A}(t)$ is given by:

$$\mathcal{A}(t) = \left(\frac{GM}{c^2 \, r} \right) \left(\frac{GM}{c^2 \, a(t)} \right) = \left(\frac{GM}{c^2 \, r} \right) \left[\frac{64}{5} \, \frac{c^3 (t_c - t)}{GM} \right]^{-1/4}. \qquad (8.7.14)$$

From the above equation one can make an order of magnitude estimate of $\mathcal{A}$. For individual masses of about a solar mass, at a typical distance of about 200 Mpc and at a GW frequency of 100 Hz, $\mathcal{A} \sim 10^{-23}$ which is also the order of magnitude of the GW amplitudes h_+ and $h_\times$. To get an idea of the time of coalescence, we may choose f_0 at the lowest end of the band-width. For two masses, each one solar mass and $f_0 = 20 Hz$, it takes about $t_c \sim 273$ seconds to coalesce. This is roughly the time the GW signal sweeps through the bandwidth of the detector. The time the signal spends in the sensitive band rises rapidly as the f_0 is reduced—in fact $\propto f_0^{-8/3}$. So it is especially important to improve the sensitivity at the lower end of the bandwidth of the detector in order to improve signal detection.

A very important observation which established the existence of GW is the Hulse-Taylor binary pulsar. The orbit of the binary decays because of loss of energy by the emission of GW and exactly as predicted by GR. In Exercise 4 we have given the steps in which the decrease in period of the orbit observed in the Hulse-Taylor binary pulsar can be derived using the foregoing analysis.

8.8 Sources

As is evident from the quadrupole formula, Eq. (8.5.15), any massive system with the quadrupole moment varying with time can generate gravitational waves. Thus, if we start spinning with our arms stretched out, we would start emitting gravitational waves! Although its amplitude will be incredibly small—impossible to detect by current (or near future) instruments. When compact stars, like black holes coalesce, they release enormous amount of energy in GW. A binary star emits gravitational waves which carry away energy from the system leading to reduction in its orbital distance as the stars fall deeper into the potential well with total energy becoming more negative. As the orbit shrinks, the orbital period reduces and the system keeps emitting increasingly stronger GW and losing energy at an even faster rate. This runaway process eventually leads to the merger of the binary with an enormous release of energy. Both the frequency and amplitude of GW keep increasing until the merger, which resembles the waveform of a bird's chirp. The quadrupole formula allows us to calculate the GW signal from inspiralling compact binaries in a fairly straightforward way to the leading order [Exercise 2]. The GW signal is called a chirp. However, this gets more complicated in the latter part of the inspiral, especially, in the last few cycles before the merger, where the energy carried away by GW becomes so large that the 'back-reaction' on the component masses and their orbital evolution requires the computation of higher order terms. One computes these terms in a perturbative expansion with the small parameter $(v/c \sim \sqrt{GM/ac^2})$. This expansion is called the post-Newtonian (PN) expansion in which the non-linearities are dealt with iteratively (Blanchet (2002); Blanchet et al. (2002)). The method involves intricate and lengthy calculations. As of writing, terms have been computed upto 4PN order. The 4PN order provides adequate accuracy in phasing for a substantial part of inspiral for the matched filtering method (discussed in Chap. 9) to be successfully employed. After the final black hole is formed but not yet settled into a stationary state, the waveform can be computed perturbatively. These are called the quasi-normal modes (Vishveshwara (1970), Chandrasekhar (2002)) or the ringdown modes which decay with characterstic times and frequencies. The full waveform including the merger, however requires numerical relativity (Baumgarte and Shapiro (2010)).

From our current knowledge of astronomy, we expect different kinds of sources of GW spanning a wide range of frequency bands, though GW astronomy can reveal more sources which may be presently unknown. The stellar mass binaries merge in the 'audible frequency band', in the range few Hz to few kHz. Bigger the binary components, smaller the final coalescence frequency. For super-massive black holes (SMBH), with masses from millions to a billion solar masses, the merger frequency is of the order of $\sim 10^{-3}$Hz. Spinning neutron stars in our galaxy, with even a small (one part in a million!) asymmetry in their mass-energy distribution, can emit detectable amount of GW. So far though GW signals from none of these sources have been detected, suggesting that mass-energy distribution in these stars is even more symmetric! Supernovae can also emit bursts of GW in the frequency range of few hundred Hz, if their explosions are predominantly asymmetric. Moreover, a collection of these sources in the distant universe can create a 'stochastic background', where we may not be able to identify the individual sources (like individual stars in a distant galaxy), but the overall energy flux from a collection of those sources can bring us information on the statistical distribution of various parameters of those astrophysical sources. A stochastic background is also created by phenomena in the very early universe, most interestingly from cosmic inflation. For a thorough review on the sources of gravitational waves see Sathyaprakash and Schutz [2009].

8.9   Effect of GW on a Ring of Test Particles

We will consider a circular ring of test particles and consider the effect of a grav-itational wave which is incident on this ring. We will compute the effect in the approximation when the wavelength of the gravitational wave is much larger than the separation between the particles. This approximation is quite reasonable when we consider large scale inteferometric gravitational wave detectors with 4 km armlengths. The distance between the test masses is few km while the typical wavelength is hundreds or thousands of kilometres. The advantage of this approximation is that we can use the geodesic deviation equation to compute the effect on the test particles. This is because the Riemann tensor is of order λ^{-2}, where λ is the wavelength of the gravitational wave. In order that the approximation be valid, the separation between the particles must be much less than λ—in the situation we are considering the radius of the circle must be much smaller than λ.

We will consider the circle to be in the (x, y) plane with the centre at the origin. The centre of the circle will be the reference point. Test particles lie on this circle initially. When a gravitational wave is incident on the test particles, the relative distance between the particles changes. We consider a generic vector ξ^α connecting the origin to a test particle on the circle at the same proper time τ as the particle at the origin. Because of the incident GW, the circle of test particles will be deformed. We compute this deformation with the help of the geodesic deviation equation.

We start with the geodesic deviation equation. Let ξ^α be the connecting vector connecting points on the geodesics at the same τ, then the geodesic equation is:

$$\frac{d^2 \xi^\alpha}{c^2 d\tau^2} = R^\alpha{}_{\mu\nu\beta} U^\mu U^\nu \xi^\beta \,. \tag{8.9.1}$$

We now write down the Riemann tensor in linearised theory of GW for convenience,

$$R_{\alpha\mu\nu\beta} = \frac{1}{2} \left(h_{\alpha\beta,\mu\nu} + h_{\mu\nu,\alpha\beta} - h_{\alpha\nu,\mu\beta} - h_{\mu\beta,\alpha\nu} \right) \,. \tag{8.9.2}$$

Taking $U^\mu = (1, 0, 0, 0)$, in the geodesic deviation equation we just require $R_{\alpha 0 0 \beta}$. We find,

$$R_{\alpha 0 0 \beta} = \frac{1}{2} \left(h_{\alpha\beta,00} + h_{00,\alpha\beta} - h_{\alpha 0,0\beta} - h_{0\beta,\alpha 0} \right) \,. \tag{8.9.3}$$

Writing the geodesic equation in the TT gauge since we have $h_{\alpha 0} = 0$ the Riemann tensor components in this gauge are simply given by:

$$R_{\alpha 0 0 \beta} = \frac{1}{2} h_{\alpha\beta,00} \,, \tag{8.9.4}$$

where we have omitted the TT superscript because we take it to be understood. The geodesic equation then becomes:

$$\frac{d^2 \xi^\alpha}{d\tau^2} = \frac{1}{2} \frac{\partial^2 h^\alpha{}_\beta}{\partial t^2} \xi^\beta \,. \tag{8.9.5}$$

(The factor of c^2 cancels on both sides.) Now $d\tau \simeq dt$ (slow motion) and α ranges only over space indices $i, j, k = 1, 2, 3$ which immediately gives us:

$$\frac{d^2\xi^i}{dt^2} = \frac{1}{2}\frac{\partial^2 h^i_k}{\partial t^2}\xi^k .$$ (8.9.6)

Since $|h_{ik}| << 1$ we expect the deformation to be small as well. Let the undeformed state be given by $\xi^i = \xi^i_0$, where ξ^i_0 is a constant vector, and the deformed state by $\xi^i = \xi^i_0 + \delta\xi^i(t)$ where we now assume $|\delta\xi^i(t)| << |\xi^i_0|$. Then we further simplify Eq. (8.9.6) to obtain:

$$\frac{d^2\delta\xi^i}{dt^2} = \frac{1}{2}\frac{\partial^2 h^i_k}{\partial t^2}\xi^k$$

$$\simeq \frac{1}{2}\frac{d^2 h^i_k}{dt^2}\xi^k_0 ,$$ (8.9.7)

since $|\delta\xi^i| << |\xi^i_0|$. The approximate solution to this equation to the first order is,

$$\delta\xi^i = \frac{1}{2}h^i_k\xi^k_0 .$$ (8.9.8)

We can now apply these results to a ring of test particles in the (x, y) plane. We take the circular ring to be of unit radius. Let a wave travelling in the z-direction be incident upon the ring. We will investigate how the ring is deformed because of the wave. For simplicity, let us consider just one polarisation, say, $h_+(t) = h_0 \cos\omega t$. The initial undeformed state is given by:

$$\xi^x = \xi^x_0 = \cos\phi, \quad \xi^y = \xi^y_0 = \sin\phi$$

and also we have $h^x_x = -h^y_y = h_+(t)$. We can immediately get the deformations from Eq. (8.9.8). We then have:

$$\delta\xi^x = \frac{1}{2}h^x_x\xi^x_0 = \frac{1}{2}h_+(t)\xi^x_0 = \frac{1}{2}h_0\cos\omega t\cos\phi ,$$

$$\delta\xi^y = \frac{1}{2}h^y_y\xi^y_0 = -\frac{1}{2}h_+(t)\xi^y_0 = -\frac{1}{2}h_0\cos\omega t\sin\phi ,$$ (8.9.9)

which then gives the result,

$$\xi^x = \xi^x_0 + \delta\xi^x = (1 + \frac{1}{2}h_0\cos\omega t)\cos\phi ,$$

$$\xi^y = \xi^y_0 + \delta\xi^y = (1 - \frac{1}{2}h_0\cos\omega t)\sin\phi .$$ (8.9.10)

Eliminating ϕ between the above relations gives:

$$\frac{(\xi^x)^2}{(1 + \frac{1}{2}h_0\cos\omega t)^2} + \frac{(\xi^y)^2}{(1 - \frac{1}{2}h_0\cos\omega t)^2} = 1 ,$$ (8.9.11)

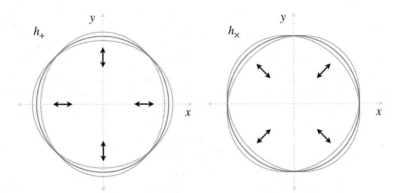

Fig. 8.2 Schematic diagram showing the effect of plane-polarised gravitational waves propagating along the z-axis incident on a circular ring of test particles in the x–y plane. The ring is deformed to an ellipse and it periodically fluctuates from elongated to compressed shape along orthogonal directions. For + polarisation these fluctuations are along the x and y axes and for $\times$ polarisation, the ellipses are rotated by 45 degrees. In usual astrophysical observations the deformations are extremely small. These have been highly exaggerated in the figure

which in general is an *ellipse* at each fixed t, as shown in Fig. 8.2. At $\omega t = 0$ the circle of test particles is deformed into an ellipse. As the time progresses to $\omega t = \pi/2$, the ellipse again becomes a circle. At $\omega t = \pi$ we again get an ellipse with the semi-major and semi-minor axes interchanged. For the cross polarisation, the ellipses are rotated by an angle of 45° with respect to the ellipses corresponding to the plus polarisation. Figure 8.2 schematically illustrates the deformations of a circular ring of test particles for both polarisations.

One may also look at a GW as a force field in the Newtonian sense. Equation (8.9.7) describes the acceleration experienced by the test particle. One can then draw the lines of force along this acceleration field. One then finds that the force field is quadrupolar.

8.10   The Response of a Laser Interferometric Detector

Consider a laser interferometric detector with arms of length L oriented along the x and y axes. A GW travelling in the Z direction is incident on the detector. We will compute its response. We first show that the response can be written as a scalar product between two tensors—one corresponding to the wave and the other corresponding to the detector. This is possible when the wavelength of the GW is much larger than L. This is called the long wavelength limit (LWL). We start with Eq. (8.9.8) and take its scalar product with the unit vector n_i and secondly take ξ_0^k along n^k and write $\xi_0^k = Ln^k$. Then the change in length δL of ξ^k is given by,

$$\delta L = \delta \xi^i n_i = \frac{1}{2} h^i_k n^k n_i L \,. \tag{8.10.1}$$

Now consider two different directions along the unit vectors $\mathbf{n}_1$ and $\mathbf{n}_2$. The response is calculated as the fractional change in the difference between the arm-lengths δL_1 and δL_2 or, that is,

$$h(t) = \frac{\delta L_1 - \delta L_2}{L} = \frac{1}{2} h_{ik}(t)(n^i_1 n^k_1 - n^i_2 n^k_2) \,, \tag{8.10.2}$$

where $h(t)$ is the response. It is a scalar.

However as it generally happens the wave may be incident from any direction with respect to the orientation of the detector. It is, therefore, useful to assign two different frames, one associated with the wave—the wave frame (X, Y, Z)- and other with the detector—the detector frame (x, y, z). The two frames in general are oriented differently with respect to each other and are related by the Euler angles (ϕ, θ, ψ). We follow the Goldstein convention for defining the Euler angles (Goldstein (1980)). We orient the wave frame so that the wave travels along the Z direction. In the TT gauge h_{ij} is a spatial tensor and is given by:

$$h_{ij}(t) = h_+(t)\, e^+_{ij} + h_\times(t)\, e^\times_{ij} \,. \tag{8.10.3}$$

where we have defined the polarisation tensors as:

$$e^+_{ij} = \hat{\mathbf{e}}_{(X)i}\hat{\mathbf{e}}_{(X)j} - \hat{\mathbf{e}}_{(Y)i}\hat{\mathbf{e}}_{(Y)j}$$
$$e^\times_{ij} = \hat{\mathbf{e}}_{(X)i}\hat{\mathbf{e}}_{(Y)j} + \hat{\mathbf{e}}_{(X)j}\hat{\mathbf{e}}_{(Y)i} \,. \tag{8.10.4}$$

Here $\mathbf{e}_{\widehat{(X)}}, \mathbf{e}_{\widehat{(Y)}}$ are unit vectors along the X, Y directions respectively and $h_{+,\times}$ are the GW amplitudes where $h_+ = h_{XX}$ and $h_\times = h_{XY}$. Note that the $e^+_{ij}, e^\times_{ij}$ defined here are purely spatial (they are in fact projections of the quantities defined in Eq. (8.4.10)).

Similarly, we choose the detector frame (x, y, z) with detector arms along the x and y axes and unit vectors $\mathbf{e}_{\widehat{(x)}}$ and $\mathbf{e}_{\widehat{(y)}}$. We define the detector tensor as:

$$D^{ij} = \frac{1}{2}\left(\hat{\mathbf{e}}^i_{(x)}\mathbf{e}_{\widehat{(x)}}{}^j - \hat{\mathbf{e}}^i_{(y)}\hat{\mathbf{e}}^j_{(y)}\right) \,, \tag{8.10.5}$$

where we have absorbed the factor of $1/2$ in the detector tensor. We could as well have absorbed this factor in the wave tensor without changing the physics. The GW signal is a scalar $h(t)$ and is given by writing it as a scalar product between the wave tensor h_{ij} and the detector tensor D^{ij}:

$$h(t) = h_{ij}(t)\, D^{ij} \,. \tag{8.10.6}$$

We can write the above scalar product as:

$$h(t) = h_+(t)\, \mathbf{e}^+_{ij}\, D^{ij} + h_\times(t)\, \mathbf{e}^\times_{ij}\, D^{ij}$$
$$\equiv h_+(t)\, F_+ + h_\times(t)\, F_\times\,,\tag{8.10.7}$$

where,

$$F_+ = \mathbf{e}^+_{ij}\, D^{ij}, \quad F_\times = \mathbf{e}^\times_{ij}\, D^{ij}\,.\tag{8.10.8}$$

The F_+ and $F_\times$ are called the *antenna pattern functions* and are functions of the Euler angles (ϕ, θ, ψ). We now compute these functions from Eq. (8.10.8) and write them first in terms of the scalar products of the unit vectors in wave and detector frames. We obtain:

$$F_+(\phi, \theta, \psi) = \mathbf{e}^+_{ij}\, D^{ij} = \frac{1}{2}[\hat{\mathbf{e}}_{(X)i}\hat{\mathbf{e}}_{(X)j} - \hat{\mathbf{e}}_{(Y)i}\hat{\mathbf{e}}_{(Y)j}]\,[\hat{\mathbf{e}}^i_{(x)}\hat{\mathbf{e}}^j_{(x)} - \hat{\mathbf{e}}^i_{(y)}\hat{\mathbf{e}}^j_{(y)}]\,,$$
$$= \frac{1}{2}[(\mathbf{e}_{\hat{(X)}} \cdot \mathbf{e}_{\hat{(x)}})^2 - (\mathbf{e}_{\hat{(X)}} \cdot \mathbf{e}_{\hat{(y)}})^2 - (\mathbf{e}_{\hat{(Y)}} \cdot \mathbf{e}_{\hat{(x)}})^2$$
$$+ (\mathbf{e}_{\hat{(Y)}} \cdot \mathbf{e}_{\hat{(y)}})^2]\,.\tag{8.10.9}$$

and,

$$F_\times(\phi, \theta, \psi) = \mathbf{e}^\times_{ij}\, D^{ij} = \frac{1}{2}[\hat{\mathbf{e}}_{(X)i}\hat{\mathbf{e}}_{(Y)j} + \hat{\mathbf{e}}_{(Y)i}\hat{\mathbf{e}}_{(X)j}]\,[\hat{\mathbf{e}}^i_{(x)}\hat{\mathbf{e}}^j_{(x)} - \hat{\mathbf{e}}^i_{(y)}\hat{\mathbf{e}}^j_{(y)}]\,,$$
$$= \left[(\mathbf{e}_{\hat{(X)}} \cdot \mathbf{e}_{\hat{(x)}})(\mathbf{e}_{\hat{(Y)}} \cdot \mathbf{e}_{\hat{(x)}}) - (\mathbf{e}_{\hat{(X)}} \cdot \mathbf{e}_{\hat{(y)}})(\mathbf{e}_{\hat{(Y)}} \cdot \mathbf{e}_{\hat{(y)}})\right]\,.\tag{8.10.10}$$

In order to express F_+ and $F_\times$ in terms of the Euler angles we need the scalar products in terms of the Euler angles. We thus need the transformation matrix between the bases $(\mathbf{e}_{\hat{(X)}}, \mathbf{e}_{\hat{(Y)}}, \mathbf{e}_{\hat{(Z)}})$ and $(\mathbf{e}_{\hat{(x)}}, \mathbf{e}_{\hat{(y)}}, \mathbf{e}_{\hat{(z)}})$. This is given by the standard matrix relating the two bases. Accordingly we obtain:

$$\begin{bmatrix} \mathbf{e}_{\hat{(X)}} \\ \mathbf{e}_{\hat{(Y)}} \\ \mathbf{e}_{\hat{(Z)}} \end{bmatrix} = \begin{bmatrix} \cos\psi\cos\phi - \cos\theta\sin\phi\sin\psi & \cos\psi\sin\phi + \cos\theta\cos\phi\sin\psi & \sin\psi\sin\theta \\ -\sin\psi\cos\phi - \cos\theta\sin\phi\cos\psi & -\sin\psi\sin\phi + \cos\theta\cos\phi\cos\psi & \cos\psi\sin\theta \\ \sin\theta\sin\phi & -\sin\theta\cos\phi & \cos\theta \end{bmatrix}$$
$$\times \begin{bmatrix} \mathbf{e}_{\hat{(x)}} \\ \mathbf{e}_{\hat{(y)}} \\ \mathbf{e}_{\hat{(z)}} \end{bmatrix}\,.\tag{8.10.11}$$

Note that the Goldstein convention for Euler angles has been used. Further, using Eqs. (8.10.9, 8.10.10, 8.10.11) gives the final result. Carrying out the calculations we find that the antenna pattern functions are (see Exercise 5):

$$F_+(\phi, \theta, \psi) = \frac{1}{2}(1 + \cos^2\theta)\cos 2\phi\cos 2\psi - \cos\theta\sin 2\phi\sin 2\psi\,,$$

$$F_\times(\phi, \theta, \psi) = -\frac{1}{2}(1 + \cos^2\theta)\cos 2\phi\sin 2\psi - \cos\theta\sin 2\phi\cos 2\psi$$
$$\tag{8.10.12}$$

These functions are the Gel'fand functions (Gelfand I. M. [1963]) or spin-weighted spherical harmonics (Goldberg et al. [1967]). See also (Dhurandhar and Tinto [1988]) where this result was first derived.

8.11   Detection of GW

Various methods have been proposed and devised to detect GW directly and indirectly at different frequencies ranging from nano-Hertz to kilo-Hertz. Two of them have succeeded so far. Enormous worldwide effort is being made to observe in these bands and thereby rapidly accelerate the field of GW astronomy. See Sathyaprakash and Schutz [2009] for a broad overview of GW astronomy.

8.11.1   Hulse-Taylor Binary Pulsar

The first indirect detection of GW was the observation of the binary pulsar system PSR B1913+16 by Russel Hulse and Joseph Taylor. Because of the emission of GW by the binary system, the orbital period of the binary reduces, thereby resulting in a slow advance of the periastron. This was precisely measured by Hulse and Taylor. The results were published in 1973 and later by Taylor and Joel M. Weisberg, over a span of few decades. The measurements matched the prediction from general relativity extremely well and provided the first indirect evidence for the emission of GW. For this discovery, Hulse and Taylor were awarded the Nobel Prize in Physics in the year 1993.[1] See Exercise 4 for a detailed calculation.

8.11.2   Bar Detectors

The first serious effort to directly detect GW was initiated by Joseph Weber at Maryland University in the 1960s. He developed highly sensitive metallic bars, that would start ringing when GWs, falling in the resonant frequency band of the detectors, were incident on the bar. The bars were equipped with highly sensitive detectors to measure any oscillations in the bars without creating any significant perturbations while sensing it. Three such bars were set up in the Maryland area, so that if the same signal is detected simultaneously, one could claim a detection of a true astrophysical signal, reducing the possibility of other instrumental noise artefacts. The more recent bar detectors equipped with cryogenics reached much better detection sensitivities, though not sufficient enough to detect GW (Aguiar 2011). Only for sources close enough, their sensitivities could be adequate, but then the probability of such events

---

[1] https://www.nobelprize.org/prizes/physics/1993/press-release/

would be very low. Nevertheless, this effort to detect tiny fluctuations was crucial for initiating the experimental program to directly detect GW.

Joe Weber is often called the father of GW astronomy.

8.11.3   Ground-Based Laser Interferometric Detectors

A realistic chance of detection of GW came with the proposal of laser interferometric detectors which are essentially Michelson interferometers, that many of us have seen in Physics labs, but these are far more sophisticated (Saulson [2017]). This was in the 1970s. Michelson interferometers are historically important in the field of relativity for they were used to establish that the velocity of light in vacuum remains the same, irrespective of the motion of the source or observer, providing a firm foundation for the principles of special relativity. Michelson and Morley devised this instrument with folded arms to increase the effective length of each arm. This principle is also applied in the modern laser interferometric detectors. The ground-based laser inteferometers typically have few kilometer arm length with Fabry-Perot cavities in each arm to effectively increase the armlength few hundred times and use techniques like 'power recycling' to increase the sensitivity even further by a factor of few tens. A schematic of the LIGO interferometers is shown in Fig. 8.3. The laser beams are enclosed in vacuum tunnels so as to eliminate the scattering of light by air as much as possible. In effect, these detectors are able to measure arm length fluctuations smaller than the size of a proton! The Fig. 8.4 shows the aerial view of one of the LIGO detectors at Livingston, Louisiana, US.

A Michelson interferometer uses a beam-splitter that splits a coherent laser beam into two nearly equal power beams along two orthogonal paths of nearly equal lengths. The beams are reflected by mirrors kept at the end of each arm which are then combined by the beam-splitter again and an interference pattern is observed on the photo-diode. The position of the fringes, which can be detected with a photo-diode, depends on the difference in light travel paths between the two arms. The Fabry-Perot design and techniques such as light recycling increase the sensitivity of the detector.

At the lower end of the detection band of frequencies, the detectors are limited mainly by seismic noise. Both passive and active isolation is used to reduce this noise. As we know from the theory of forced harmonic oscillators, a suspended pendulum driven by ground noise reaching the point of suspension, reduces the amplitude noise by a factor of $1/f^2$, for frequencies f much larger than the pendulum's resonant frequency. This can be easily seen as follows. Let a simple harmonic oscillator with a forcing term be described by the equation:

$$\ddot{x} + 2\gamma\dot{x} + \omega_0^2 x = F(t). \tag{8.11.1}$$

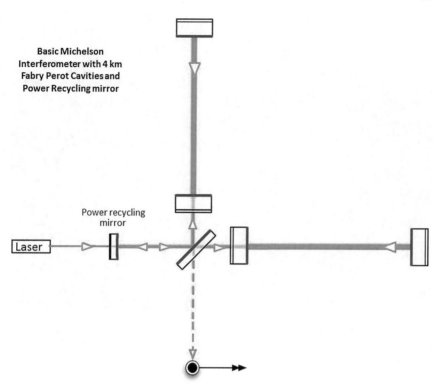

**Basic Michelson
Interferometer with 4 km
Fabry Perot Cavities and
Power Recycling mirror**

Power recycling
mirror

Laser

Fig. 8.3 A basic configuration of the LIGO interferometer with Fabry-Perot cavities in each arm. Besides the basic design, the power recycling mirror is also shown. The actual configuration is far more elaborate. Credits: Caltech/MIT/LIGO Lab

Here $x(t)$ represents the displacement, the dots represent time derivatives, the γ term describes the damping, ω_0 is the resonant frequency of the oscillator and $F(t)$ is the forcing term. Typically, the resonant frequency is a fraction of a Hertz and we measure GW at frequencies greater than 10 Hz. Equation (8.11.1) can be solved by taking Fourier transforms. In the limit of $\omega \gg \omega_0$ and γ, we see that $\tilde{x}(\omega) \propto 1/\omega^2$. Thus a suspended pendulum can act as a passive seismic isolator. LIGO uses multi-stage suspensions (with up to four stages) for different optical components of the detector. Moreover, ground noise sensed by seismometers, accelerometers etc. can be fed back at different stages of suspension to actively reduce the noise. Since the target is to measure very small distance fluctuations and also because the compact binaries have most of their orbital cycles in the lower frequencies (why?), low frequency sensitivity is crucial for increasing the signal-to-noise ratio (see Chap. 9 for discussion).

The mid-frequencies (~ 100Hz) are dominated by thermal noise because the mirrors and the suspension are at room temperature. The thermal energy of $k_B T/2$ per degree of freedom, where k_B is the Boltzmann constant and T is the temperature in Kelvin, excites different oscillation modes in the suspension and the mirror. This

Fig. 8.4 The figure shows an aerial view of the Ligo observatory at Livingston, Louisiana, US. One arm of the interferometer 4 km long is fully visible in the picture. Also the central building which houses the laser and other laser optics is visible. Credits: Caltech/MIT/LIGO Lab

noise can overwhelm a astrophysical signal. One way to reduce this noise in detection band of frequencies is by choosing the appropriate materials for the construction of the mirrors. The noise then tends to concentrate in frequencies far away from the frequencies of interest (Hough and Rowan [2000]).

The high frequencies are dominated by shot noise arising from photon counting at the photo-detector. It arises essentially from the uncertainty principle in quantum mechanics:

$$\Delta N \, \Delta \phi \, \sim \, 1, \tag{8.11.2}$$

where ΔN is the standard deviation in the mean number of photons $\bar{N}$ emitted by the laser in some duration τ (the period of the GW wave at the frequency one is measuring) and $\Delta \phi$ the uncertainty in the phase which manifests itself as shot noise. The noise is Poissonian in nature. In the Poisson distribution the variance is equal to mean—in this case $\bar{N}$ and so the standard deviation $\Delta N \sim \sqrt{\bar{N}}$ and therefore $\Delta \phi \sim 1/\sqrt{\bar{N}}$. Thus $\Delta \phi$ or the shot noise reduces as the photon number flux $\bar{N}/\tau$ or as the laser power $P = \bar{N}\hbar\omega_L/\tau$ is increased. Here ω_L is the frequency of the laser. Generally for the current design of the detectors, the requisite senstivity results when, $\Delta \phi \sim 10^{-9}$, 10^{-10} if we wish to measure GW with strain sensitivity $h \sim 10^{-22}$. For a GW frequency of 100 Hz we therefore need $\bar{N} \sim 10^{18}$ photons arriving in 0.01 s. For $\omega_L \sim 2\pi \times 3 \times 10^{14}$ radians/sec (Nd-Yag laser), this computation gives $P \sim 100$ W. However, for the first generation LIGO, the laser power was limited to only few Watts.

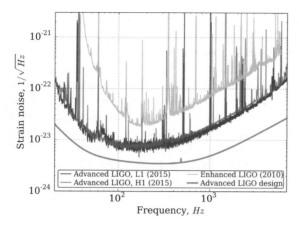

Fig. 8.5 The figure shows the sensitivity curves of the LIGO observatories at Hanford (H) and Livingston (L). At low frequencies seismic noise dominates, at mid frequencies it is the thermal noise and at high frequencies the shot noise takes over. Also spikes are seen—these are the resonances of the suspensions. The vertical axis is $\sqrt{S_h(f)}$ (see Eq. 9.3.14 of Chapter 9) or the strain per $\sqrt{\text{Hz}}$. The blue and red curves are the sensitivity curves in 2015 for Advanced LIGO for the Livingston and the Hanford respectively. The black curve shows the design sensitivity curve. Reprinted figure with permission from Martynov et al. [2016], Phys. Rev. D93, 112004 (2016). Copyright (2016) by the American Physical Society

Here is where the idea of power recycling is important. Power recycling effectively multiplies this base power by few tens which then enhances the power to the required level.

However, increasing the laser power also creates more fluctuations in the radiation pressure exerted by the laser beam on the freely hanging mirrors. Thus, the laser power cannot be increased indefinitely without increasing the masses of the mirrors. Increasing the masses reduces the acceleration caused by the radiation pressure. However, increasing the mirror mass requires significant changes in the other design parameters. This is an example of how closely knit all parts of the detectors are and why it takes years to decades to improve the sensitivity of the instrument. In fact, it is the cutting-edge technology and decades of research which made it possible to directly detect gravitational waves, 100 years after they were predicted!

The LIGO sensitivity curves are shown in Fig. 8.5. The sensitivity curves are shown for enhanced LIGO (green) in 2010 and in 2015 for Advanced LIGO for the Hanford (red) and Livingston (blue) LIGO detectors. The black curve shows the design sensitivity curve.

We have discussed here only three main sources of noise, which essentially determine the sensitivity curve. There remain many other sources of noise which plague the detector and we refer the reader to the relevant literature (Adhikari 2014; Hough and Rowan 2000).

The first generation detectors, LIGO (US), Virgo (Italy) and TAMA (Japan), were indeed able to measure tiny fluctuations, though the probability of detection with those detectors was small. The second generation detectors, Advanced LIGO[2] and Advanced Virgo[3], reached adequate sensitivity to detect gravitational waves. Recently, another detector, namely, KAGRA[4] in Japan, has joined the network. LIGO-India[5] (Indo-US collaboration), a laser interferometric detector built on Indian soil, is expected to join the network by the end of this decade. The main advantage of this extended network is that the longer baselines will enable GW astronomers to localise the sources with much more precision. Further the number of detections will increase, thereby increasing the chances for observing electromagnetic radiation from the binary neutron star mergers which is a treasure trove for science. The second generation detectors probe only a tiny fraction of the sources in the audible band. The proposed third generation detectors, Einstein Telescope[6] and Cosmic Explorer[7], will be able to detect essentially the most distant binary mergers.

8.11.4   Space-Based Detectors

Sensitivities of the ground-based detectors are plagued by a host of noise sources at low frequencies, such as the seismic noise, gravity gradient noise etc. It will be very difficult to observe GW with such detectors say below few Hertz. However astrophysically, we know that there exist very interesting GW sources at low frequencies in the milli-Hertz and deci-Hertz frequency range. In order to observe such low frequency sources, the answer is to go to space.

Super-massive black holes (SMBH) merge in the mHz frequency band [Exercise 6]. Also, there are other sources such as extreme-mass ratio inspirals (EMRI), where a relatively small stellar mass black-hole/neutron star spirals into an SMBH. Such sources also emit GW in the low frequency band. These exciting sources can be observed with the planned space-based observatories like the Laser Interferometer Space Antenna (LISA) joint space mission of the European Space Agency (ESA) and NASA or the TianQin mission by China, targetted for launch in 2030s. Moreover, these missions will be able to observe and predict when a stellar-mass binary is going to merge from weeks to years in advance. This knowledge will be very useful to the ground-based detectors for preparing in advance to observe the event and plan for electromagnetic follow up. The proposed deci-Hertz detector Deci-hertz Interferometer Gravitational wave Observatory (DECIGO) will also be able to provide such alerts. Most interestingly, this band may provide a window to observe the stochastic background generated in the early universe with minimal contamination from astrophysical sources.

---

[2] https://www.ligo.caltech.edu

[3] http://www.virgo.infn.it

[4] https://gwcenter.icrr.u-tokyo.ac.jp/en/

[5] https://www.ligo-india.in

[6] http://www.et-gw.eu

[7] https://cosmicexplorer.org

The proposed space-based detectors are very long interferometers, though their designs can vary significantly. In the space based detectors, the test masses are "freely falling" enclosed in spacecraft. The spacecraft equipped with sophisticated sensors and micro-thruster technology follow the test masses creating a "drag-free" environment. LISA is proposed to have an armlength of 2.5×10^6 km, while DECIGO's armlength may be $\sim 10^3$ km. The arm length somewhat relates to the target observation range of GW wavelengths. For more information see Kembhavi and Khare [2020].

8.11.5 Pulsar Timing Array (PTA)

Pulsars are spinning neutron stars with electromagnetic radiation in the form of beams directed out from their magnetic polar regions. These beams are rotating and visible from earth if they are so directed that the earth lies in the path swept by the beam. As observed from earth, they are seen as repeated pulses of electromagnetic waves as the pulsar spins. These pulses arrive at regular intervals, the fractional change in the interval can be less than one part in a billion. However, when a GW passes by, it will create a small jitter in the time of arrival of the pulse. One typically tracks several pulsars. One then expects to find a correlation between these jitters produced by the GW. By precisely tracking these correlated fluctuations in pulse timing from a well studied array of pulsars, it is possible to detect the presence of GW and even locate the astrophysical source. This is the principle of pulsar timing array (PTA). Several radio antennas across the world are working together to utilise this technique to observe GW in the nano-Hertz band. Inspiralling SMBH and the stochastic background from the early universe are the primary target sources for PTA. With the upcoming square kilometer array (SKA) radio telescope, PTA has a good chance of reaching adequate sensitivity to observe the target sources.

8.11.6 Cosmic Microwave Background (CMB)

The universe is expanding. It was much smaller in the past and its temperature was much higher. It was so dense that photons were constantly colliding with electrons, and the photons could not travel freely. Subsequently, as the universe cooled due to expansion, the electrons and protons combined to form neutral atoms and the photons could start propagating freely. This is called the recombination epoch. It occurred when the ambient temperature of the universe was about $T \sim 3000°$ K corresponding to an energy $kT \sim 0.3$ eV. However, the ionisation energy of the hydrogen atom is ~ 13.6 eV. The universe had to cool further than is implied by this energy because of various reasons (see Padmanabhan [1993]). According to the standard model of cosmology, this happened when the universe was about 3,80,000 years old. These photons come to us almost isotropically and this is the earliest signal that we receive in the form of electromagnetic waves, known as, the Cosmic Microwave Background (CMB). The radiation at the present epoch has a Planckian spectrum corresponding to a temperature of $\sim 2.7°$ K. Though CMB is highly isotropic, there are small anisotropies in its temperature and polarisation that carry the past history of the universe. A certain component of the polarisation anisotropy, called the 'B-mode', can contain the signature of primordial gravitational wave background. However, gases in our own galaxy can contaminate this signature and disentangling the two is challenging. Detecting the signature of primordial GW background is the primary target for the proposed CMB missions.

8.12   Direct Detection of Gravitational Waves

A spectacular prediction of general relativity is GW. This prediction has been vindicated recently. The first direct detection of gravitational waves occurred on September 14, 2015 (LIGO Scientific Collaboration et al. 2016). The Advanced LIGO detectors were performing engineering runs before the observation runs could formally start. A binary black hole merger event was detected. Researchers spent next few months investigating if an instrumental or terrestrial artefact could mimic this signature, but they failed to find any. They estimated that the chance of noise versus signal as predicted by general relativity with two LIGO detectors about 3000 km away, was about one in 65, 000 years. That is, the probability that this detection was not of astrophysical origin was extremely small, thus confirming the detection of a true astrophysical signal with high statistical significance.

GW sources are named following the convention GWyymmdd, where "yymmdd" is the date of the detection in UTC. When more than one source is detected on the same day, a time-stamp is added to the name. The first detected event is thus called GW150914. It was a merger of a 35 $M_\odot$ and a 30 $M_\odot$ black holes which were about 400 Mpc away. The event lasted in the detector's frequency band for about 0.2 sec. The final black hole had a mass of 62 $M_\odot$ and the total energy released in the merger was equivalent to about 3 $M_\odot c^2$.

GW150914 was the most exciting event because it was the first directly observed event! Some may argue that the most exciting event detected so far was the binary neutron star merger GW170817, because it was accompanied by electromagnetic counterparts covering a wide range of frequencies from γ-rays to radio. It was observed by a large number of electromagnetic telescopes spanning several frequency bands. This was possible because the Virgo detector was also sensitive enough at that time, enabling localisation of the event in a relatively small region of the sky. This single event had an immense scientific outcome. It showed that short gamma ray bursts are associated with binary neutron star mergers, and that heavy metals are formed in such events without which it was difficult to explain the abundance of those elements in the universe. And in the context of cosmology, we could directly estimate the distance to the binary merger from GW observations without the customary distance ladder. Further, if a host galaxy could be identified with the merger event and its red-shift measured from electromagnetic observations, it gave an independent estimate of the Hubble constant.

About ninety binary mergers have been detected by the GW detectors so far, including interesting sources like GW190521 (LIGO Scientific Collaboration et al. [2021b]). This event establishes the existence of black holes with masses greater than 100 solar masses. A large number of detections are expected in the coming years, which will help us address fundamental questions like how many black hole binaries are there in the universe, where they form, how they are distributed in the universe and many more. With the upgraded and more advanced detectors, we expect to detect gravitational waves from different kinds of sources like the supernovae, milli-second pulsars, stochastic sources etc. These discoveries can bring dramatic changes in our understanding of the universe. Gravitational wave astronomy is just born; it has had an eventful beginning! It promises an exciting future ahead.

Exercises

1. Use Eqs. (8.4.6), (8.4.7) and (8.4.8) to solve explicitly for B_μ in terms of $A_{\mu\nu}$ and k^μ.
2. Compute the GW amplitudes for an observer in a arbitrary direction at an angle ι with the orbital spin axis of the binary system. Assume that the orbit of the binary to be circular.

   (a) Fill in the algebra between Eqs. (8.7.6) and (8.7.7) in the relevant derivation.
   (b) Use the projection operator P^i_j defined in Sect. 8.4.2 to obtain the same result.

3. Consider the general case of two point masses $m_1 \neq m_2$ and total mass M, in circular orbit at a separation distance a and orbital frequency Ω. Find the expressions for the I_{ij}, L_{GW}, t_c and finally the GW amplitudes h_+ and $h_\times$. Carry out the steps as outlined below following essentially the derivations 18–20.

   (a) Compute the quadrupole tensor I_{ij}. Choose the orbit in the $(x - y)$ plane. Let $a_1 = (m_1/M)a$, $a_2 = (m_2/M)a$ and $a = a_1 + a_2$ with the centre of mass at the origin. Show that,

   $$I = \frac{1}{2}\mu a^2 \begin{bmatrix} \cos 2\phi(t) & \sin 2\phi(t) & 0 \\ \sin 2\phi(t) & -\cos 2\phi(t) & 0 \\ 0 & 0 & 0 \end{bmatrix}, \qquad (8.12.1)$$

   where $\phi(t) = \int \Omega(t)dt$ and $\Omega(t)$ adiabatically evolves with time. Here $\mu = m_1 m_2/M$ is the reduced mass.

   (b) Show that the GW luminosity is:

   $$L_{GW} = \frac{32}{5}\frac{G}{c^5}\mu^2 a^4 \Omega^6 = \frac{32}{5}\frac{G^4 M^3 \mu^2}{c^5 a^5}. \qquad (8.12.2)$$

The second equality has been obtained from Kepler's law.

(c) From the total energy:

$$E = -\frac{1}{2}\frac{G\mu M}{a},$$ (8.12.3)

obtain the equation for the decaying separation $a(t)$ as:

$$a^3\frac{da}{dt} = -\frac{64}{5}\frac{G^3\mu M^2}{c^5},$$ (8.12.4)

and hence obtain the coalescence time t_c,

$$t_c = \frac{5}{256\eta}\frac{c^5}{G^3 M^3}a_0^4 = \frac{5}{256\eta}\frac{GM}{c^3}\left(\frac{\pi GM f_0}{c^3}\right)^{-8/3},$$ (8.12.5)

where the initial separation is a_0 and corresponding GW frequency f_0. Note that the GW frequency $f = \Omega/\pi$ which is twice the orbital frequency. Also $\eta = \mu/M$ is the reduced mass ratio.

(d) Define the chirp mass $\mathcal{M}$ as $\mathcal{M} = [\mu M^{2/3}]^{3/5}$. This quantity makes the expressions look simpler. By integrating Eq. (8.12.4) from t to t_c, obtain the amplitude and phase as

$$\mathcal{A}(t) = \frac{G\mathcal{M}}{c^2 r}\left[\frac{c^3(t_c - t)}{5G\mathcal{M}}\right]^{-\frac{1}{4}}$$

$$\phi(t) = \phi_c - \left[\frac{c^3(t_c - t)}{5G\mathcal{M}}\right]^{5/8}.$$ (8.12.6)

(e) From the above expressions, derive the GW amplitudes,

$$h_+ = +\frac{1}{2}\mathcal{A}(t)\,(1 + \cos^2\iota)\,\cos[\,2\,\phi(t)\,],$$

$$h_\times = +\mathcal{A}(t)\,\cos\iota\,\sin[\,2\,\phi(t)\,],$$ (8.12.7)

where $\mathcal{A}(t)$, $\phi(t)$ are given by Eq. (8.12.6).

4. Hulse-Taylor binary pulsar: The binary pulsar $1913 + 16$ was observed by the radio-astronomers Hulse and Taylor for few decades. They found a decay in orbit of the pulsars; the period of the orbit decreased at the rate of about 70 μs per year. This decrease in the period was attributed to loss of energy from the emission of GW, thus observationally establishing the existence of GW. In this exercise we obtain this period decrease from the formulae in this chapter by carrying out the following steps:

(a) The period of the binary orbit P is defined by $P = 2\pi/\Omega$. Using Kepler's law and Eq. (8.7.9) and taking logarithmic derivatives with respect to time obtain

$$\frac{\dot{E}}{E} = -\frac{\dot{a}}{a} = \frac{2}{3}\frac{\dot{\Omega}}{\Omega} = -\frac{2}{3}\frac{\dot{P}}{P},$$

where the dot over a quantity represents the time derivative.
(b) From Eqs. (8.7.8) and (8.7.9) show that for a circular orbit:

$$\frac{\dot{P}}{P} = -\frac{24}{5}\frac{G^3 M^3}{c^5 a^4}. \tag{8.12.8}$$

But the Hulse-Taylor binary pulsar is not in a circular orbit; the orbit is elliptical with eccentricity $e \sim 0.617$. Therefore, the expression for $\dot{P}/P$ in Eq. (8.12.8) must be modified; it must be multiplied by the factor $f(e)$ (Peters and Mathews 1963), where,

$$f(e) = \frac{1 + \frac{73}{24}e^2 + \frac{37}{96}e^4}{(1 - e^2)^{7/2}}.$$

(c) For the Hulse-Taylor binary pulsar, assume the following values for the relevant quantities:

$$P \sim 27907 \text{ s} \approx 7.75 \text{ hours}, \quad e \sim 0.617 \implies f(e) \sim 11.84,$$

$$M \approx 2.8 \, M_\odot, \quad a \sim 1.944 \times 10^6 \text{ km},$$

to arrive at $\dot{P} \sim -2.4 \times 10^{-12}$. Hence deduce that in one year the period decreases by $\Delta P \sim 72 \, \mu s$.

5. Fill in the calculations and use Eqs. (8.10.9, 8.10.10) and Eq. (8.10.11) to derive the antenna pattern functions in terms of the Euler angles as in Eq. (8.10.12).
6. Show that the orbital frequency of a binary black hole of total mass M in its last stable circular orbit with a separation of $6GM/c^2$, reduces as the mass M increases. Estimate this frequency for a SMBH of mass $10^6 \, M_\odot$ by applying Kepler's law for Newtonian orbits.

Chapter 9
Gravitational Wave Data Analysis

9.1 The Task of a GW Data Analyst

Ground based as well as space based gravitational wave (GW) detectors will be continuously spewing out data. Since gravity couples to matter weakly, the gravitational wave signals are expected to be weak; the noise generally overwhelms the signal and so the signal will be deeply buried in the noise. The task of the GW data analyst is to extract the weak signal embedded in the noise by (i) saying whether a signal is present or absent, and (ii) secondly, if present, then to measure its parameters. The signal is expressed by a tiny metric perturbation h_{ik} where $|h_{ik}| << 1$. This perturbation may be due to compact binary star coalescences, supernovae, asymmetric spinning neutron stars, stochastic GW background, or some unknown source so far not discovered electromagnetically. We denote a typical component of the perturbation simply by h; we disregard the indices which essentially provide directional information. The metric (and the perturbation) is a dimensionless quantity and in typical situations $h \sim 10^{-23}$ or smaller. For a GW source, h can be estimated from the well-known Landau-Lifschitz quadrupole formula derived in the last chapter, namely, Eq. (8.5.15). The GW amplitude h is related to the second time derivative of the quadrupole moment (this quantity has the dimensions of energy) of the source and hence from the quadrupole formula:

$$h \sim \frac{4}{R} \frac{G}{c^4} E^{\text{kinetic}}_{\text{non-spherical}} , \qquad (9.1.1)$$

where R is the distance to the source and $E^{\text{kinetic}}_{\text{non-spherical}}$ is the kinetic energy in the *non-spherical* motion of the source. From Eq. (9.1.1) the h can be estimated. If we consider $E^{\text{kinetic}}_{\text{non-spherical}}/c^2$ a fraction of a solar mass and the distance to the source ranging from the galactic scale of tens of kpc to cosmological distances of Gpc, then

---

The original version of this chapter was revised: Belated corrections in equations have been updated. The correction to this chapter is available at https://doi.org/10.1007/978-3-030-92335-8_10

S. Dhurandhar and S. Mitra, *General Relativity and Gravitational Waves*,
UNITEXT for Physics, https://doi.org/10.1007/978-3-030-92335-8_9

169

h ranges from 10^{-17} to 10^{-23}. These numbers then set the scale for the sensitivities at which the GW detectors must operate. GW is detected by measuring the change in arm-length of a laser interferometric detector which is observed as a phase shift on a photo-diode. If the change in the arm-length L of the detector is δL, then, in the long wavelength limit (LWL), when $\lambda_{GW} >> L$, which applies to ground based detectors,

$$\delta L \sim hL . \tag{9.1.2}$$

The average noise amplitude in current detectors on the other hand is of the same order or few orders larger than the signal amplitude. Typically, the signal is deeply embedded in the noise. Therefore the data analyst needs to extract the GW signal from the noisy data. If the signal stands well above the noise, then the task is easy, because then it is hard for the noise to mimic the signal. On the other hand, if the signal is weak and is buried in the noise as is usually the case, in order to detect and/or measure the signal, we need foreknowledge of the signal. The more we know about the signal, the better the chances of our success. For instance, we might know the form of the signal. But even if we know the form of the signal or which amounts to the knowledge of the family of signal waveforms—for example, compact binary coalescences—the signals may depend on several parameters, which are unknown in the case of a specific signal. Then the task of the analyst is to search the parameter space of the family of signals and find the signal with the specific parameters. We will see how this can be achieved by employing digital filters, more specifically, we will discuss in some detail the matched filtering technique.

Another important aspect of the detection problem arises because the data contains noise which is only described statistically. Let the signal be denoted by $h(t)$ where t denotes the time and the noise by $n(t)$. The noise $n(t)$ is a stochastic process (Papoulis and Pillai [2002]) and at each time t it is a random variable. We will sample the data with discrete set of samples, and hence the data will be a collection of a finite set of samples. If the data are just noise and if the N noise samples are denoted by $n(t_k)$, $k = 0, 1, ..., N - 1$, then we are dealing with a N dimensional random vector and its multivariate probability distribution. Therefore, any claim of detection must be a statistical statement. One can never be sure of a 100% detection. Because there is noise in the data it is possible that the noise conspires to appear like a signal. One can only assign probabilities to the detection process and provide confidence levels for claiming a detection. This is the best that we can do. We will see in this chapter how we can achieve this goal in an optimal way.

In this chapter, we will start with time series data, and characterise noise in terms of its power spectral density (PSD), colour and probability distribution. We will then describe matched filtering and discuss its optimal properties. The Neyman-Pearson criterion is appropriate in the context of GW detection and the Neyman-Pearson lemma in the context of hypothesis testing will be discussed. We will then discuss the likelihood ratio and maximum likelihood. Finally we will argue how geometry provides an elegant framework for GW data analysis with the concepts of manifolds, metric etc. playing a key role.

9.2 Time Series Data and Its DFT

As remarked before the data are sampled at closely placed time instants and so a discrete form of the output is available from a detector. We will assume that the data are already calibrated using the detector transfer functions and the time series directly contains the GW amplitude $h(t)$. This is called the h of t channel of the detector (the detector data is recorded in a large number of channels, for example, seismic, etc.).

Let us start with a data segment $[0, T]$ which may be part of the full data say from $t = 0$ to $t = T_{\text{obs}}$. For example, in the recent runs of LIGO and Virgo, when searching for compact coalescing binaries, T could be of the order of several minutes and T_{obs} could be several weeks during which the data quality was acceptable. Although our data consists of real numbers, there is no harm at this stage by considering a slightly general case of complex valued data. Accordingly, consider a complex valued function $h(t)$ defined on the interval $[0, T]$. Thus $h : [0, T] \longrightarrow C$. We divide $[0, T]$ into N equal sub-intervals of length Δ and take time samples $t_k = k\Delta$, $k = 0, 1, 2, \ldots, N - 1$. Therefore the samples of the function h are $h_k = h(t_k) = h(k\Delta)$ which are N in number. This is called *uniform* sampling. There are applications in which non-uniform sampling is preferred, but for our current purpose uniform sampling suffices. Then we have the relation $T = N \Delta$.

The set of samples can be written as a column vector $\mathbf{h} \in C^N$ (we denote vectors by boldface letters):

$$\mathbf{h} = \begin{bmatrix} h_0 \\ h_1 \\ \vdots \\ h_{N-1} \end{bmatrix}. \tag{9.2.1}$$

The time series data is real however and so we will be dealing actually with real vectors and so $\mathbf{h} \in \mathcal{R}^N$. We will later put additional structure on $\mathcal{R}^N$ by defining a scalar product on $\mathcal{R}^N$. We will then have a Hilbert space which we will denote by $\mathcal{D}$—the space of data vectors. Then $\mathbf{h} \in \mathcal{D}$.

An important tool in processing the data is the Fourier transform. Since the data are discretely sampled and the number of samples are finite, the Discrete Fourier Transform or DFT in short is applicable. The following equation demonstrates how the DFT naturally arises from the Fourier expansion of $h(t)$ over $[0, T]$. We have,

$$\tilde{h}_n = \int_0^T h(t)e^{-2\pi i n t/T}\, dt \quad -\infty < n < \infty$$

$$\simeq \Delta \sum_{k=0}^{N-1} h(t_k)e^{-2\pi i n t_k/T} = \Delta \sum_{k=0}^{N-1} h_k e^{-2\pi i n k/N} \equiv \Delta \tilde{H}_n. \tag{9.2.2}$$

Here $\tilde{h}_n$ is the Fourier coefficient and in fact Eq. (9.2.2) defines the DFT $\tilde{H}_n$ (Press et al. [2007]). The integral is written approximately as a Riemann sum. Writing out the DFT explicitly:

$$\tilde{H}_n = \sum_{k=0}^{N-1} h_k e^{-2\pi i nk/N} , \qquad (9.2.3)$$

where $\tilde{H}_n$ is called the discrete Fourier transform (DFT) of h_k. Also $\tilde{H}_n$ is complex because $\mathbf{h}$ has been expanded on a complex basis $e^{-2\pi i nk/N}$, $k = 0, 1, ..., N - 1$. However, unlike the Fourier expansion which has infinitely many Fourier coefficients, here we have only N quantities; only N successive values of n are needed to fully determine the DFT. The usual choices are, from 0 to $N - 1$, or equivalently $-N/2$ to $N/2 - 1$. This also follows from the fact that $\mathbf{h}$ is an N-dimensional vector and also from the fact that the DFT possesses the circularity property, $\tilde{H}_{n+N} = \tilde{H}_n$. The samples therefore span a N dimensional space C^N and the DFT can be looked upon as the change of components of the vector $\mathbf{h}$ from the time domain basis to the Fourier basis. The DFT relation can be inverted and then we get the Inverse DFT or IDFT. It is given by:

$$h_k = \frac{1}{N} \sum_n \tilde{H}_n e^{2\pi i kn/N} . \qquad (9.2.4)$$

We just write n under the summation sign to portray the flexibility of its range. Thus we can easily switch from the time domain representation of h to a Fourier one and vice-versa.

In order to establish certain theoretical results, we will often switch from the discrete representation of data to the continuous and vice-versa. In the continuous case, the sums are replaced by integrals. Our approach here will be intuitive rather than rigorous.

9.3   Characterisation of Noise

9.3.1   Random Variables and their Distributions

We start with a simple example. A function X takes on values $x_1, x_2, ..., x_N$, which are generically denoted by x_i . In any single instance we do not know what value X will take; it can take any one of the values x_i. But in a large number of trials, we can predict the results fairly reliably; we can predict the relative frequencies with which X will take the value x_i fairly accurately. As the number of trials tends to infinity, the relative frequencies approach definite values, say p_i. These are the probabilities with which X takes the value x_i. We may write:

$$P\{X = x_i\} = p_i , \quad i = 1, 2, ..., N . \qquad (9.3.1)$$

X is called a random variable (often abbreviated by r. v.) and Eq. (9.3.1) describes its distribution. The probabilities satisfy, $p_i > 0$ and $p_1 + p_2 + ... + p_N = 1$. More generally, a r.v. X is function from a sample space Ω to $\mathcal{R}$ the set of real numbers. In the above example, Ω is the set of integers from 1 to N. Subsets of Ω are called events and they are assigned probabilities. Often we will deal with continuous distributions where X can take any value $x \in \mathcal{R}$. Then the probabilities of an r. v. are prescribed via a distribution function $F_X(x)$. We denote the event $X^{-1}((-\infty, x])$ simply by $\{X \leq x\}$ which is a subset of the sample space $\mathcal{R}$. Then the distribution function of X is defined as:

$$F_X(x) = P\{X \leq x\}, \tag{9.3.2}$$

where P denotes the probability of the event under consideration. We will require such functions when we talk about false alarm and detection probabilities. F_X is a non-decereasing function of x and having the property, $F_X(-\infty) = 0$, $F_X(\infty) = 1$. Usually we deal with the derivative of F_X which is called the *probability density function* or in short just 'pdf'. We have:

$$p_X(x) = \frac{dF_X}{dx}, \tag{9.3.3}$$

where we have denoted the pdf by p_X. From the above properties of F_X, we must have $p_X(x) \geq 0$ and

$$\int_{-\infty}^{\infty} p_X(x)dx = 1. \tag{9.3.4}$$

The interpretation is that the probability that X takes a value in the interval $[x, x + dx]$ is $p_X(x)\, dx$.

An important distribution is the Gaussian or the normal distribution. The standard normal distribution is defined by the pdf:

$$p(x) = \frac{1}{\sqrt{2\pi}} e^{-x^2/2}, \qquad -\infty < x < \infty. \tag{9.3.5}$$

Its mean is $\mu = 0$ zero and variance $\sigma^2 = 1$. They are the first and second moments of the distribution and are given via the equations:

$$\mu = \langle x \rangle = \int_{-\infty}^{\infty} x\, p(x)\, dx = 0$$

$$\sigma^2 = \langle x^2 \rangle - \mu^2 = \int_{-\infty}^{\infty} x^2\, p(x)\, dx = 1. \tag{9.3.6}$$

The standard normal distribution of X is written as $X \sim N(0, 1)$, where N stands for normal; 0 is the mean and 1 the variance. The angular brackets denote ensemble average. The term ensemble will be explained in the next subsection in the concrete context of GW detectors. From the standard normal distribution, we can easily transform to a r.v. Y by setting $X = (Y - \mu)/\sigma$. Then Y has a normal distribution with mean μ and variance σ^2. We write $Y \sim N(\mu, \sigma^2)$.

However, the situation we are in is more general. We have to deal collectively with several samples of the data, each of which is a r.v. and these r.v.s could be correlated. We are then dealing with random vectors and their multivariate probability distributions. Noise in fact is a stochastic process, any N samples of which can be considered as a random vector. We have a data segment $[0, T]$ and it is sampled uniformly with N points—we have the samples of the data $x(t_k = k\Delta) = x_k$, $k = 0, 1, 2, ...N - 1$. We can look upon the samples x_k as components of a N-dimensional vector, say a column vector, which we denote by $\mathbf{x}$. Then each of the x_k are r.v.s and $\mathbf{x}$ is a random vector distributed according to a N dimensional multivariate distribution. The multivariate distribution is described by the distribution function $F(x_0, x_1, ..., x_{N-1})$ or the pdf $p(x_0, x_1, ..., x_{N-1})$ which can be compactly written as $F(\mathbf{x})$ or $p(\mathbf{x})$ respectively. The interpretation is that each component x_i of $\mathbf{x}$ *jointly* takes the value between x_i and $x_i + dx_i$ with the probability $p(x_0, x_1, ..., x_{N-1})\, dx_0\, dx_1\, ...\, dx_{N-1}$. Analogously, the multivariate distribution function is defined as $F(x_0, x_1, .., x_{N-1}) = P\{X_0 \le x_0, X_1 \le x_1, ..., X_{N-1} \le x_{N-1}\}$.

9.3.2 Stationary, Coloured and Gaussian Noise

We apply the above considerations to the sampled noise denoted by the random vector $\mathbf{n}$. The noise then is described by the pdf $p(\mathbf{n})$. Also we will assume that the noise has zero mean. We can always do this for the detector noise by subtracting out the DC component. We then have each $\langle n_i \rangle = 0$ or in vector form $\langle \mathbf{n} \rangle = 0$. In the continuous representation of data, we will consider the noise to be a zero mean stochastic process—at each time t, $n(t)$ is a r.v. At each time t we write,

$$\langle n(t) \rangle = 0 \,, \tag{9.3.7}$$

where the angular brackets, as before, denote *ensemble* average. One imagines a large number of identical detectors producing data. Then at each instant of time t, we take the average of $n(t)$ over the ensemble of the detectors. More specifically, consider M identical detectors labelled by $I = 1, 2, ..., M$ and their corresponding noise vectors $\mathbf{n}_I$, then, for each time instant t, the ensemble average is defined as:

$$\langle n(t) \rangle = \frac{1}{M} \sum_{I=1}^{M} n_I(t) \,. \tag{9.3.8}$$

In the Fourier space $\tilde{n}(f)$ is also a r. v. because it is the Fourier transform of a r.v. We have,

$$\tilde{n}(f) = \int n(t)\, e^{-2\pi i f t}\, dt \,. \tag{9.3.9}$$

In the discrete domain it is a sum of r.v.s multiplied by phases which change with different time instants—the frequency f is fixed for a particular $\tilde{n}(f)$. Since $\tilde{n}(f)$ is a r. v. we can compute its mean:

$$\langle \tilde{n}(f) \rangle = \left\langle \int n(t) e^{-2\pi i f t} dt \right\rangle = \int \langle n(t) \rangle e^{-2\pi i f t} dt \equiv 0. \qquad (9.3.10)$$

We find that the mean of the r.v. $\tilde{n}(f)$ vanishes.

Now consider the second moment. We characterise the noise by its autocorrelation function K as the covariance of the noise at two time instants t and t'. It is defined as the ensemble average of the product $n(t)n(t')$ as follows (note that the mean of $n(t)$ is zero):

$$\langle n(t)n(t') \rangle = K(t, t'). \qquad (9.3.11)$$

Clearly from the definition $K(t, t')$ is symmetric in its arguments, namely, $K(t, t') = K(t', t)$. We will normally put restrictions on this function and hence the noise. The noise is called stationary if for any N samples, the distribution function or the pdf is invariant under time translations. That is,

$$F[n(t_0), n(t_1), ..., n(t_{N-1})] = F[n(t_0 + \tau), n(t_1 + \tau), ..., n(t_{N-1} + \tau)], \qquad (9.3.12)$$

for all time translations τ. We will normally put a weaker restriction on the noise requiring only the first two moments to be invariant under time translations. That is:

$$\langle n(t) \rangle = \langle n(t + \tau) \rangle, \quad \langle n(t)n(t') \rangle = \langle n(t + \tau)n(t' + \tau) \rangle, \qquad (9.3.13)$$

for all time translations τ. Such a stochastic process is called *wide sense stationary* or in short WSS. We will assume that the noise is WSS. In practice this is not the case, but if the signals are of short duration, then the noise in the neighbourhood of the signal can be considered to be approximately WSS.

When the noise is WSS, we have $K(t, t') = K(t + \tau, t' + \tau)$ for all τ. We may now specifically choose $\tau = -t'$, then $K(t, t') = K(t - t', 0)$ or K becomes a function of only one variable $t - t'$. Without any cause for confusion, we denote it by the same symbol as $K(t - t')$. Physically, this means that the mean and the covariance between noise samples do not depend on absolute time but only on their time difference.

Because of the time translation symmetry, in Fourier space we have a simple representation of the autocorrelation function. In the Fourier space, we have:

$$\langle \, \tilde{n}(f) \tilde{n}^\star(f') \rangle = \int \int dt \, dt' \, \langle n(t) \, n(t') \rangle \, e^{-2\pi i f t + 2\pi i f' t'}$$

$$= \int \int dt \, dt' \, K(t - t') \, e^{-2\pi i f t + 2\pi i f' t'}$$

$$= \int \int dt \, d\tau \, K(\tau) \, e^{-2\pi i f t + 2\pi i f'(t+\tau)}$$

$$= \int d\tau \, K(\tau) \, e^{2\pi i f \tau} \times \delta(f - f')$$

$$= S(f) \, \delta(f - f') \, . \tag{9.3.14}$$

Here the star denotes the complex conjugate of a quantity. Note that $\tilde{n}(f)$ is a complex r.v. for each frequency f. We have also changed variables by defining $t' = t + \tau$ in the integrals and written $S(f)$ for the Fourier transform of K. We will justify this notation a little later in the text. Because of stationarity we get a delta function $\delta(f - f')$ in the frequency space. Since $K(\tau) = K(-\tau)$, that is, it is an even function, its Fourier transform $S(f)$ is real. Also since from Eq. (9.3.14), $S(f) = \langle |\tilde{n}(f)|^2 \rangle$; it is an average of positive quantities, and hence it is positive definite. It is called the power spectral density or in short PSD of the noise. $S(f)$ is also an even function, namely, $S(-f) = S(f)$ and it can be folded on itself and defined only for $f \geq 0$. Then it is called the one sided PSD denoted by $S_{\text{one sided}}(f)$. Eq. (9.3.14) becomes:

$$\langle \, \tilde{n}(f) \tilde{n}^\star(f') \rangle = \frac{1}{2} \, S_{\text{one sided}}(f) \, \delta(f - f') \, . \tag{9.3.15}$$

In almost all of the literature it is the one sided PSD that is used. If the time is measured in seconds, $S(f)$ has dimensions of Hz^{-1}. In the literature usually the subscript "h" is added to S and $S(f)$ is written as $S_h(f)$ to signify that the PSD refers to the data channel which contains the signal $h(t)$. But here, we will continue to write the PSD as $S(f)$ in order to avoid clutter, with no cause for confusion.

We can characterise the noise as white or coloured. If $S(f) = S_0$ a constant, then the noise is called white; otherwise it is coloured. In case of white noise, the autocorrelation function becomes,

$$K(\tau) = S_0 \int df \, e^{2\pi i f \tau} \equiv S_0 \, \delta(\tau) \, . \tag{9.3.16}$$

So then we have the relations:

$$\langle n(t) n(t') \rangle = S_0 \, \delta(t - t'), \qquad \langle \tilde{n}(f) \tilde{n}^\star(f) \rangle = S_0 \, \delta(f - f') \, . \tag{9.3.17}$$

The noise samples at different times are uncorrelated however small the difference $t - t'$ is. This is clearly not realistic since any detector has a non-zero response time. If one takes samples separated by times less than the response time of the detector, they are bound to be correlated. White noise therefore is an idealisation. Noise can be considered white in some limited band which may be of interest where the function $S(f)$ is more or less constant.

We now define Gaussian white noise. Consider a data segment $[0, T]$ uniformly sampled with N samples and sampling interval Δ; we then have $t_k = k\Delta$, $k = 0, 1, ..., N - 1$. We integrate $\langle n(t)n(t_k)\rangle$ over a sample width:

$$\int_{t_k-\Delta/2}^{t_k+\Delta/2} dt \, \langle n(t)\, n(t_k)\rangle = S_0 \int_{t_k-\Delta/2}^{t_k+\Delta/2} dt \, \delta(t - t_k). \tag{9.3.18}$$

The LHS gives $\langle n_k^2\rangle \, \Delta$, while the RHS is just S_0. Thus we have the relation:

$$\langle n_k^2\rangle = \frac{S_0}{\Delta} \equiv \sigma_\Delta^2. \tag{9.3.19}$$

Now let each n_k be distributed Gaussian with mean zero and variance σ_Δ^2. But because the different noise samples are mutually statistically independent, the joint pdf of the random vector $\mathbf{n}$ is the product of the individual pdfs of the r.v.s n_k. Thus the pdf for the noise vector $\mathbf{n}^T = (n_0, n_1, ..., n_{N-1})$ is:

$$p(\mathbf{n}) = \frac{1}{(\sqrt{2\pi}\sigma_\Delta)^N} e^{-\frac{1}{2}\frac{|\mathbf{n}|^2}{\sigma_\Delta^2}}, \tag{9.3.20}$$

where $|\mathbf{n}|^2 = \mathbf{n}^T\mathbf{n} = \sum_{k=0}^{N-1} n_k^2$ which is in fact the square of the Euclidean norm $\| \mathbf{n} \|_E^2$. This is the pdf of white Gaussian noise.

The discussion can be extended in a straight forward way to coloured Gaussian noise. We define the covariance matrix of the noise vector as:

$$C_{ik} = K(t_i - t_k) = K[(i - k)\,\Delta]. \tag{9.3.21}$$

Then the pdf of the noise r.v. $\mathbf{n}$ is:

$$p(\mathbf{n}) = \frac{1}{(2\pi)^{N/2}\sqrt{\det C}} e^{-\frac{1}{2}\mathbf{n}^T C^{-1}\mathbf{n}}. \tag{9.3.22}$$

This is the pdf of coloured Gaussian noise. The covariance matrix is obtained via $C = \langle \mathbf{n}\,\mathbf{n}^T\rangle$ by direct integration. See Exercise 1.

We may define the matrix G as the inverse of C, then in the exponent of the above pdf given in Eq. (9.3.22) we get the quantity $G_{ik}n_i n_k$. This motivates the definition of a scalar product. We define a scalar product of two real data vectors (they have real components in the time domain) $\mathbf{x}$ and $\mathbf{y}$ as:

$$(\mathbf{x}, \mathbf{y}) = G_{ik}x_i y_k. \tag{9.3.23}$$

Then we can write:

$$p(\mathbf{n}) = \frac{(\det G)^{1/2}}{(2\pi)^{N/2}} e^{-\frac{1}{2}(\mathbf{n},\mathbf{n})}. \tag{9.3.24}$$

If we call the set of data trains as $\mathcal{D}$ which is essentially $\mathcal{R}^N$, then with the scalar product so defined, $\mathcal{D}$ is a Hilbert space. For the special case of white noise the metric G_{ik} reduces to δ_{ik}, the Kronecker delta, then $\mathcal{D}$ is a N-dimensional Euclidean space.

In the Fourier space in the case of WSS noise, the covariance matrix is diagonalised. In the discrete form, $C_{ik} = S(f_i)\,\delta_{ik}$. Since the metric G is the inverse of the covariance matrix C, $G_{ik} \longrightarrow 1/S(f)$. In fact, the equation (9.3.23) in the continuous domain can be written as,

$$(\mathbf{x},\ \mathbf{y}) = \int_{-\infty}^{\infty} df \, \frac{\tilde{x}^\star(f)\,\tilde{y}(f)}{S(f)} \equiv 4\,\mathrm{Re}\left[\int_0^{\infty} df \, \frac{\tilde{x}^\star(f)\,\tilde{y}(f)}{S_{\text{one sided}}(f)}\right]. \qquad (9.3.25)$$

The vector space of data trains $\mathcal{D}$ equipped with the above defined scalar product becomes a Hilbert space. The norm of a vector $\mathbf{x}$ is then defined as $\| \mathbf{x} \| = +\sqrt{(\mathbf{x}, \mathbf{x})}$. The norm defines a metric on $\mathcal{D}$ and hence a topology. Topological considerations become important when defining boundaries of subsets of $\mathcal{D}$. These will be required when setting thresholds for assiging statistical significance to signals.

The noise in a real detector is coloured and a detector has a bandwidth in which it is sensitive to GW signals. Any real detector has a non-zero response time in which it reacts to a signal. This response time determines the upper limit frequency of the frequency band in which the detector is sensitive.

9.4 Matched Filtering

When a known signal is embedded in the noise, matched filtering is the optimal method for extracting the signal from the noise. A known signal is that when the waveform of the signal can be determined. It may happen that we may be able to compute the form of the signal but it may depend on several parameters. Then in that case we must search through the space of parameters. Also we will discuss here in what sense the matched filter is optimal. We first start with a simple example and then generalise.

9.4.1 A Simple Example

We take the example of a sinusoidal signal. A sinusoidal signal has a simple form described by,

$$h(t) = A\cos(2\pi f_0 t + \phi)\,, \qquad (9.4.1)$$

where f_0 is a constant frequency, A the amplitude and ϕ a constant phase. We do not know apriori the parameters A, f_0 and ϕ. These must be determined from the matched filtering operation. The signal is embedded in the noise $n(t)$. We take the noise to be additive, then the data $x(t)$ are given by,

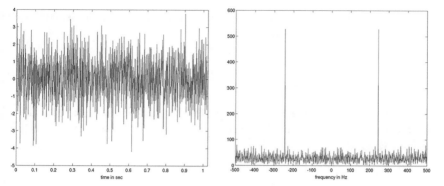

Fig. 9.1 The figure on the left shows time series data. A sinusoidal signal of smaller amplitude is embedded in noise of comparatively larger amplitude. The signal is not directly visible to the eye. In the figure on the right, the statistic $C(f)$ is plotted versus the frequency f. Two peaks are seen at $\pm f_0$ in the plot. In this figure we have taken $f_0 = 256$ Hz and $\phi = 0$. From the position and the height of the peaks we can estimate f_0 and A respectively

$$x(t) = n(t) + h(t). \qquad (9.4.2)$$

In Fig. 9.1 we have taken the amplitude of the sinusoid to be smaller than the typical amplitude of the noise so that the signal is not directly visible to the eye in the raw data. We now apply the matched filter to this data. In this case the matched filter is also a sinusoid and is easily implemented efficiently by taking the Fourier transform of the data by employing the FFT algorithm. We compute the statistic $C(f)$:

$$C(f) = |\tilde{x}(f)|, \quad \tilde{x}(f) = \int_{-\infty}^{\infty} dt\, x(t)\, e^{-2\pi i f t}, \qquad (9.4.3)$$

where $\tilde{x}(f)$ is the Fourier transform of $x(t)$ as defined above. The statistic $C(f)$ is just the modulus of the Fourier transform. Two peaks are seen in the Fig. 9.1 in the figure on the right, occuring at $\pm f_0$ because $\cos 2\pi f_0 t = (e^{-2\pi i f_0 t} + e^{2\pi i f_0 t})/2$. If we set a threshold sufficiently high so that it is unlikely that the noise crosses this threshold we could claim a detection of the signal with some probability or statistical significance. We will discuss the significance of detection later in this chapter. Apart from detection, we also recover parameters of the signal from the statistic. The frequency is determined from position of the peaks and the amplitude from their heights. Additionally the phase can also be determined from the phase of the complex Fourier transform.

9.4.2 The Matched Filter and Signal to Noise Ratio

Consider a data segment $[0, T]$ and a known signal $h(t)$. This signal is buried in the noise whose auto correlation function is $K(t, t')$. In general we have not put any restrictions of stationarity at this stage. Then the matched filter $q(t)$ is defined as the solution of the integral equation:

$$\int_0^T K(t, t')\, q(t')\, dt' = h(t),\tag{9.4.4}$$

where $h(t)$ denotes the signal. If we sample the data, then in the discrete form, the above equation becomes a matrix equation. The matrix must be inverted to obtain the filter vector $\mathbf{q}$.

If the noise is WSS, we have the equation:

$$\int_0^T K(t - t')\, q(t')\, dt' = h(t).\tag{9.4.5}$$

The integral is a convolution and we can readily obtain q by going over to the frequency domain by taking Fourier transforms. Indeed in the Fourier space the above equation reduces to $\tilde{q}(f)\, S(f) = \tilde{h}(f)$ and since h is known, we immediately have the solution for the matched filter $\mathbf{q}$:

$$\tilde{q}(f) = \frac{\tilde{h}(f)}{S(f)}.\tag{9.4.6}$$

Figure 9.2 shows the plot of the correlation when a matched filter is applied to the data containing the GW binary signal. In the middle section of the plot, it is difficult to see the signal in the data, as it is overwhelmed by the noise. In the bottom section of the plot, a peak in the correlation is observed at the time of arrival of the signal. In the example we gave in the last subsection, the Fourier transform is the matched filter for a sinusoidal signal.

We now define the signal-to-noise ratio (SNR) ρ of a given statistic and apply these considerations to the matched filter. In Fourier space we have:

$$\tilde{x}(f) = \tilde{h}(f) + \tilde{n}(f).\tag{9.4.7}$$

Let us apply the filter $\tilde{q}(f)$ to the data. Then we obtain the statistic c which is given by,

$$c = \int \tilde{q}^\star(f)\, \tilde{x}(f)\, df.\tag{9.4.8}$$

The integrals go from $-\infty$ to ∞ which we have not explicitly written. We note that c is real for real filters and real data and also note that c is also a r.v. since it depends on the data $\mathbf{x}$ which is a r.v. because the data contains noise. The average SNR denoted by $\langle \rho \rangle$ is the ratio of the mean of c to its standard deviation. The mean of c is denoted by $\mu_c = \langle c \rangle$ and it is given by:

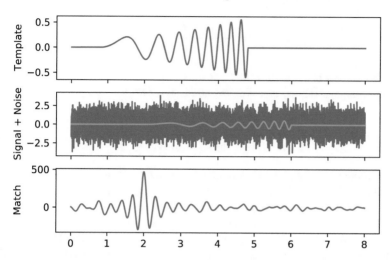

Fig. 9.2 The basic operation of matched filtering is shown in the figure. The top panel shows the template of the signal. The middle panel shows the signal buried in the noise. The bottom panel shows the output of the matched filter. The template slides along the time axis to generate the matched filter output, which peaks at the time of arrival of the signal

$$
\begin{aligned}
\mu_c = \langle c \rangle &= \left\langle \int \tilde{q}^*(f)\, \tilde{x}(f)\, df \right\rangle \\
&= \left\langle \int \tilde{q}^*(f)\, \tilde{h}(f)\, df \right\rangle + \int \tilde{q}^*(f)\, \tilde{n}(f)\, df \\
&= \int \tilde{q}^*(f)\, \tilde{h}(f)\, df .
\end{aligned}
\tag{9.4.9}
$$

The last line follows from the fact that the second term is zero because the mean of the noise is zero, that is, from Eq. (9.3.10) we have $\langle \tilde{n}(f) \rangle = 0$. Also the first term is deterministic. The variance of c, say σ_c^2, is obtained from taking the expected value of the filter operating on the noise only. This is because $\sigma_c^2 = \langle (c - \langle c \rangle)^2 \rangle$ and therefore only the noise term survives. Thus we have:

$$
\begin{aligned}
\sigma_c^2 &= \left\langle \int \tilde{q}(f)\, \tilde{n}^*(f)\, df \int \tilde{q}^*(f')\, \tilde{n}(f')\, df' \right\rangle \\
&= \int df\, df'\, \tilde{q}(f)\, \tilde{q}^*(f')\, \langle \tilde{n}^*(f)\, \tilde{n}(f') \rangle \\
&= \int df\, df'\, \tilde{q}(f)\, \tilde{q}^*(f')\, S(f)\, \delta(f - f') \\
&= \int df\, |\tilde{q}(f)|^2\, S(f) .
\end{aligned}
\tag{9.4.10}
$$

Then the SNR is given by:

$$\rho = \frac{c}{\sigma_c} = \frac{\int \tilde{q}^*(f)\, \tilde{x}(f)\, df}{[\int df\, |\tilde{q}(f)|^2\, S(f)]^{1/2}} \, . \qquad (9.4.11)$$

Note that since **x** is a r. v. so is the SNR ρ.

The average SNR, $\langle \rho \rangle$ is just μ_c/σ_c and is given by,

$$\langle \rho \rangle = \frac{\mu_c}{\sigma_c} = \frac{\int \tilde{q}^*(f)\, \tilde{h}(f)\, df}{[\int df\, |\tilde{q}(f)|^2\, S(f)]^{1/2}} \, . \qquad (9.4.12)$$

In the literature sometimes $\langle \rho \rangle$ is referred to as the SNR instead of ρ. In this book we will refer to $\langle \rho \rangle$ as the average or mean SNR and ρ as simply the SNR.

We observe that $\langle \rho \rangle$ depends on the filter q we apply, which is itself a function, while the average SNR $\langle \rho \rangle$ is a real number. Thus $\langle \rho \rangle$ maps the function q to a real number $\langle \rho \rangle(q)$ - it is called a *functional*. Secondly, we also observe that scaling q, that is, $q \longrightarrow Aq$, where A is a constant, does not change $\langle \rho \rangle$. Thus $\langle \rho \rangle$ is invariant under scalings of the filter q. We can use this fact to simplify the algebra in the calculations. Accordingly, we choose A so that,

$$\int df\, |\tilde{q}(f)|^2\, S(f) = 1 \, . \qquad (9.4.13)$$

We have then normalised the filter. For the normalised filter the SNR is defined as,

$$\rho = \int \tilde{q}^*(f)\, \tilde{x}(f)\, df \, , \qquad (9.4.14)$$

and the average SNR:

$$\langle \rho \rangle = \int \tilde{q}^*(f)\, \tilde{h}(f)\, df \, . \qquad (9.4.15)$$

Below we give an example of the matched filter for the Newtonian waveform of two inspiraling compact binaries. We take the more general waveform given in Exercise 3 of Chap. 8. In the derivation below, we use this waveform. However, we put $t_c = \phi_c = 0$ for simplicity. For the general case of t_c and ϕ_c the result obtained below must be multiplied by the factor $e^{-2\pi i f t_c + 2i\phi_c}$. The matched filter is then obtained by dividing $\tilde{h}(f)$ by the PSD $S(f)$ as given in Eq. (9.4.6).

Derivation 21 Fourier transform of the Newtonian chirp waveform

Here we compute the Fourier transform of the waveform with the help of the stationary phase approximation (SPA) (Mathews and Walker [1970]). We start with the waveform given in Exercise 3 of Chap. 8. We put $t_c = \phi_c = 0$. We therefore take the waveform as:

$$h(t) = \mathcal{A}(t)\cos[2\phi(t)] \equiv A[-t/\alpha]^{-1/4}\cos[(-t/\alpha)^{5/8}],$$

where $A = G\mathcal{M}/r$ and $\alpha = 5G\mathcal{M}/c^3$. The Fourier transform is given by:

$$\tilde{h}(f) = \int_{t_a}^{0} dt\, h(t)\, e^{-2\pi i f t},$$

where $t_a < 0$, the time of arrival of the signal. We can write the integrand by splitting the cosine into two exponentials. Also we compute the Fourier transform for $\omega = 2\pi f > 0$:

$$\tilde{h}(f) = \frac{1}{2}\int_{t_a}^{0} dt\, \mathcal{A}(t)\, e^{i\psi(t)},$$

where,

$$\psi(t) = -\omega t - 2(-t/\alpha)^{5/8}.$$

We have written only one of the exponentials which gives a non-zero answer for which the total phase $\psi(t)$ has a stationary point in the integration range. The other exponential gives negligible contribution because it rapidly oscillates. The amplitude $\mathcal{A}(t)$ is slowly varying and can be pulled outside the integral at the stationary point which we denote by $t_0(\omega)$. By setting $\dot{\psi} = 0$ we solve for $t_0(\omega)$ to obtain:

$$t_0(\omega) = -(5/4)^{8/3}\frac{(\alpha\omega)^{-5/3}}{\omega}.$$

The stationary phase approximation yields:

$$\tilde{h}(\omega) \approx \frac{1}{2}\mathcal{A}(t_0(\omega))\, e^{i\psi(t_0(\omega))}\sqrt{\frac{2\pi}{\ddot{\psi}(t_0(\omega))}}\, e^{i\pi/4}.$$

We obtain after some algebra:

$$\mathcal{A}(t_0(\omega)) = 4\frac{G\mathcal{M}}{c^2 r}\left(\frac{\pi G\mathcal{M}f}{c^3}\right)^{2/3},\quad \psi(t_0(\omega)) = -\frac{3}{128}\left[\frac{\pi G\mathcal{M}f}{c^3}\right]^{-5/3}.$$

$$\ddot{\psi}(t_0(\omega)) = \frac{48}{5}\left[\frac{\pi G\mathcal{M}f}{c^3}\right]^{5/3}(2\pi f)^2$$

Putting all the pieces together one arrives at:

$$\tilde{h}(f) = \sqrt{\frac{5\pi}{24}}\frac{G^2\mathcal{M}^2}{c^5 r}\left(\frac{\pi G\mathcal{M}f}{c^3}\right)^{-7/6} e^{i\left[\frac{\pi}{4} - \frac{3}{128}\left(\frac{\pi G\mathcal{M}f}{c^3}\right)^{-5/3}\right]}$$

9.4.3   The Optimality of the Matched Filter

We now prove that the matched filter produces the highest average SNR among all linear filters. This is in one way the matched filter is optimal. This statement is true irrespective of whether the noise is Gaussian or not.

Let q be any function satisfying the normalisation condition $\int df \, |\tilde{q}(f)|^2 S(f) = 1$. It is mapped to the number $\langle \rho \rangle (q)$ via Eq. (9.4.15). We now show that $\langle \rho \rangle (q)$ is maximised when q is the matched filter given by Eq. (9.4.6).

This can be most conveniently achieved by invoking the Schwarz inequality. We set,

$$u(f) = \tilde{q}(f)\sqrt{S(f)}, \qquad v(f) = \frac{\tilde{h}(f)}{\sqrt{S(f)}}, \tag{9.4.16}$$

and write,

$$u \cdot v \equiv \int u^*(f) \, v(f) \, df = \int \tilde{q}^*(f) \, \tilde{h}(f) \, df . \tag{9.4.17}$$

where we have invoked the usual L^2 scalar product. Note that this scalar product is different from the scalar product introduced in Eq. (9.3.25), in which the inverse of the PSD appears as a weight function. In the usual L^2 scalar product in Eq. (9.4.17) there is no weight function or equivalently the weight function is unity.

Then we have from the Schwarz inequality,

$$|u \cdot v| \leq \| u \| \ \| v \| \equiv \| v \| ,$$

$$= \left| \int df \, \frac{|\tilde{h}(f)|^2}{S(f)} \right|^{\frac{1}{2}} , \tag{9.4.18}$$

because $\| u \| = 1$ by virtue of Eq. (9.4.13) which says $\| u \|^2 = 1$. Maximisation is attained when one of the vectors is a scalar multiple of the other or when the vectors "point in the same direction". That occurs when $u(f) = A v(f)$ where A is a constant. From Eqs. (9.4.16) and (9.4.17) we obtain,

$$\tilde{q}(f) \sqrt{S(f)} = A \frac{\tilde{h}(f)}{\sqrt{S(f)}} . \tag{9.4.19}$$

The matched filter is determined upto a scaling. Thus,

$$\tilde{q}(f) = A \frac{\tilde{h}(f)}{S(f)} . \tag{9.4.20}$$

The constant A can be fixed by requiring the filter $\tilde{q}(f)$ to have unit norm. We have shown that the q given by Eq. (9.4.20) maximises the average SNR $\langle \rho \rangle$ among all *linear filters*. Then the maximum average SNR $\langle \rho \rangle$ is given by $\| v \|$ as given in Eq. (9.4.18).

The same result can be obtained by maximising the functional $\langle \rho \rangle (q)$ with respect to q by using functional derivatives. See Exercise 2.

9.5 Statistical Significance of Detection

If the right matched filter is applied to the data containing a signal, the maximum of the output of the matched filter, will be on the average, the SNR ρ. In order to decide whether there is in fact a signal embedded in the data, the SNR should cross a preset threshold. It is possible that the noise masquerades as a signal, but in general, typically its chance of doing so reduces if the threshold is set sufficiently high. But what is meant by sufficiently high? We need to quantify these notions precisely. There is already considerable literature addressing these problems in the statistics of signal detection (See Helstrom [1968]).

In this section we will describe hypothesis testing and restrict ourselves to the binary hyposthesis. In this context, we will discuss the Neyman-Pearson criterion and lemma, likelihood ratio etc. In the next section we will follow up with composite hypothesis, maximum likelihood and also present an elegant geometrical framework in which these questions can be well framed.

9.5.1 Binary hypothesis testing

Given a data vector $\mathbf{x}$ we must decide whether a specific or known signal is present or absent in the data. The data vector $\mathbf{x}$ is a vector random variable with a multivariate probability distribution. The probability distributions will differ depending on whether there is signal present in the data or not. Given $\mathbf{x}$ we must decide which of these distributions the vector $\mathbf{x}$ belongs to. When there is no signal in the data, then the data vector is just the noise $\mathbf{x} = \mathbf{n}$; while if there is a signal $\mathbf{h}$ in the data then, the data vector is $\mathbf{x} = \mathbf{h} + \mathbf{n}$. The hypotheses are called H_0 and H_1 respectively. The multivariate probability distributions are denoted by $p_0(\mathbf{x})$ when the hypothesis is H_0 and $p_1(\mathbf{x})$ when the hypothesis is H_1. Given $\mathbf{x}$ we must decide between the hypotheses H_0 and H_1. Since we have assumed additive noise we also have $p_1(\mathbf{x}) = p_0(\mathbf{x} - \mathbf{h})$.

To decide between the hypotheses, we device a test. A test amounts to partitioning the range of $\mathbf{x}$ into two disjoint regions $\mathcal{D}_0$ and $\mathcal{D}_1$, that is if the range of $\mathbf{x}$ is $\mathcal{D}$, then $\mathcal{D} = \mathcal{D}_0 + \mathcal{D}_1$, where the plus sign denotes disjoint union. Thus if $\mathbf{x} \in \mathcal{D}_1$, the signal is present or H_1 is true or else $\mathbf{x} \in \mathcal{D}_0$, the signal is absent. Generally, the region $\mathcal{D}_1$ is far away from the origin of $\mathcal{D}$ while $\mathcal{D}_0$ surrounds the origin. This is because we have assumed the mean of the noise to be zero; the distribution of the noise vector $\mathbf{n}$ is clustered around the origin. The boundary of $\mathcal{D}_1$ namely, $\partial \mathcal{D}_1$ is called the decision surface D. Usually for well behaved probability distributions like

Gaussian for example, D is a smooth $N - 1$ dimensional hypersurface of $\mathcal{D}$. So now the problem of deciding whether H_0 or H_1 is true amounts to determining the region $\mathcal{D}_1$ according to some specific criteria. In order to do this we define the false alarm and detection probabilities as follows:

$$P_F(\mathcal{D}_1) = \int_{\mathcal{D}_1} p_0(\mathbf{x}) \, d^N x \,,$$

$$P_D(\mathcal{D}_1) = \int_{\mathcal{D}_1} p_1(\mathbf{x}) \, d^N x \,. \tag{9.5.1}$$

P_F is called the false alarm probability and P_D is called the detection probability. Further $1 - P_D$ is called the false dismissal probability while $1 - P_F$ is called the confidence. The false alarm probability is the probability of the noise masquerading as a signal, that is, choosing H_1 when H_0 is in fact true. While the false dismissal probability is the probability of saying there is no signal in the data, when there is in fact a signal, that is choosing H_0 over H_1 when H_1 in fact is true.

9.5.2 Neyman-Pearson Criterion

So then how does one decide between H_0 and H_1? One way is to assign costs to each type of decision. Then we have a cost matrix c_{ij}, where c_{ij} is the cost incurred by choosing hypothesis H_i, when H_j is true. A criterion could be to minimise the average cost given the prior probabilities for the hypotheses H_0 and H_1. This is called the Bayes criterion. However, let us consider the problem more simply. In GWDA, it is difficult to assign costs in any meaningful way; there are no tangible costs to saying there is a GW signal, when there is no signal and vice-versa. The criterion that seems to be most appropriate in this situation is the Neyman-Pearson criterion. Here we fix the false alarm probability to some small value that we can tolerate and maximise the detection probability with respect to this constraint. This is a well posed problem and the solution is given by the Neyman-Pearson lemma which then determines the region $\mathcal{D}_1$.

When events are rare and the costs of making mistakes is unquantifiable, the Neyman-Pearson criterion is appropriate. We fix the false alarm probability $P_F(\mathcal{D}_1) = \alpha$ to a small number that can be tolerated. It is fixed essentially by the event rate that one expects. If one expects 40 or 50 events (say compact binary coalescences) in a year, then one may fix the false alarm rate to one or two false alarms per year. We may afford to make one or two mistakes among 50 true events.

The Neyman-Pearson criterion states:

Maximise the detection probability $P_D(\mathcal{D}_1)$ subject to $P_F(\mathcal{D}_1) = \alpha$.

The question is how do we find $\mathcal{D}_1$? The answer is supplied by the Neyman-Pearson lemma (which we do not prove here. See Helstrom [1968]). It essentially provides a prescription to find $\mathcal{D}_1$, which is the following:

Define the likelihood ratio:

$$\Lambda(\mathbf{x}) = \frac{p_1(\mathbf{x})}{p_0(\mathbf{x})}, \tag{9.5.2}$$

then the regions which maximise P_D are of the form $\mathcal{D}_1(\Lambda_0) = \{\mathbf{x}/\Lambda(\mathbf{x}) \geq \Lambda_0\}$ where Λ_0 is a parameter.

Now we choose the parameter Λ_0 such that $P_F(\mathcal{D}_1(\Lambda_0)) = \alpha$. This equation fixes Λ_0. Λ_0 is called the *threshold*. The optimal region $\mathcal{D}_1$ in the sense of Neyman-Pearson is then determined as also is the decision surface D which is then given by the equation $\Lambda(\mathbf{x}) = \Lambda_0$.

We illustrate the situation for Gaussian noise. In general, the noise could be coloured. Then the pdfs for the signal absent and signal present cases are respectively the following:

$$p_0(\mathbf{x}) = A_N \, e^{-\frac{1}{2}G_{ik}x_i x_k},$$
$$p_1(\mathbf{x}) = A_N \, e^{-\frac{1}{2}G_{ik}(x_i - h_i)(x_k - h_k)}. \tag{9.5.3}$$

Here $\mathbf{h}$ is the N dimensional signal vector with components $h_i, i = 0, 1, 2, \ldots N - 1$, A_N is the normalisation constant which normalises the pdfs (it is the same for both pdfs), G_{ik} is the inverse of the covariance matrix and Einstein summation convention is used for repeated indices. We may now compute $\Lambda(\mathbf{x})$ which is the ratio of the two pdfs. But it is more convenient to compute the log likelihood ratio, firstly because of the exponentials occurring in the form of the pdfs and secondly because the logarithm is a monotonic function of its argument. We can then set the threshold on the log likelihood ratio instead of the likelihood. We then have:

$$\ln \Lambda(\mathbf{x}) = G_{ik}x_i s_k - \frac{1}{2}G_{ik}h_i h_k$$
$$\equiv (\mathbf{x}, \mathbf{h}) - \frac{1}{2} \| \mathbf{h} \|^2. \tag{9.5.4}$$

The monotonicity of the functions involved allows us to translate the equation, $\Lambda(\mathbf{x}) \geq \Lambda_0$ to $\rho \geq \rho_0$ where,

$$\rho = (\mathbf{x}, \mathbf{h}) = G_{ik}x_i h_k. \tag{9.5.5}$$

Thus we just need to compute the scalar product of the data vector with the signal vector. We may write,

$$\rho = q_i x_i, \quad q_i = G_{ik}h_k, \tag{9.5.6}$$

where we recognise that q_i is just the matched filter!

Thus in Gaussian noise we have shown that the matched filter is also optimal in the Neyman-Pearson sense—it maximises the detection probability for a given false alarm probability. The decision surface is just $\mathbf{q} \cdot \mathbf{x} = \rho_0$ which is a hyperplane in $\mathcal{D}$.

9.6 Composite Hypothesis and Maximum Likelihood

9.6.1 The Signal Manifold

Although the discussion in the previous section is simple and straightforward to apply, we are not in this situation. In general we do not have a *specific* known signal buried in the data. But what we do know, in fact, is the form of the signal. Generally, we know the family of signals, say $h(t; \lambda^\alpha)$ which depend on a number of parameters λ^α, $\alpha = 1, 2, \ldots p$. For example in the case compact coalescing binary sources, the signal depends on the individual masses, m_1, m_2, the amplitude A, spins and other kinematical parameters like the time of coalescence, initial phase etc. *Apriori*, we do not know the values of these parameters. In fact, apart from deciding whether there is a signal in the data or not, the further problem is to estimate the parameters of the signal. From astrophysical considerations or otherwise, we may have some idea about the ranges of the parameters. We must then look for the signal in this range. All these parameters constitute the λ^α in this case and they span some subset $U \subset \mathcal{R}^p$. We collectively denote the λ^α by the vector $\vec{\lambda}$. Thus $\vec{\lambda} \in U$, where U is the domain of the vector parameter $\vec{\lambda}$. Also we will denote the class of signals by $\mathbf{h}(\vec{\lambda})$, where now $\mathbf{h}(\vec{\lambda}) \in \mathcal{D}$, the space of data trains. Thus we have a mapping $\psi : U \longrightarrow \mathcal{D}$ where $\vec{\lambda} \in U$ is mapped to $\mathbf{h}(\vec{\lambda}) \in \mathcal{D}$. The image of U under this mapping is a p dimensional manifold—a submanifold of $\mathcal{D}$. This manifold we call the *signal* manifold and denote it by $\mathcal{S}$. The signal parameters λ^α can be regarded as coordinates on $\mathcal{S}$. This is why we denote them with the superscript α. In the differential geometric language we have the chart U with the homeomorphism ψ. Generally, the GW waveforms can be considered to be smooth functions (C^∞) of the parameters $\vec{\lambda}$. This means that $\mathcal{S}$ is a smooth differentiable manifold with coordinates λ^α in the patch U. We will later define a metric on $\mathcal{S}$, thus making $\mathcal{S}$ into a Riemannian manifold.

9.6.2 Maximum Likelihood

The situation we are in is the following: we do not have a single unknown signal, but we have data in which there may be a signal present whose form is known but not the signal itself. On the other hand the data may not contain any signal. In the event of a signal being present we know that it belongs to a family of signals which may depend on several parameters; however, we do not know its parameters. We are now dealing with a composite hypotheses:

- H_0: No signal present, that is, $\mathbf{x} = \mathbf{n}$, noise only and the corresponding pdf is $p_0(\mathbf{x})$.
- H_1: One among the family of signals $\mathbf{h}(\vec{\lambda})$ is present, that is, $\mathbf{x} = \mathbf{n} + \mathbf{h}(\vec{\lambda})$, the corresponding pdf being now denoted by $p_1(\mathbf{x}; \vec{\lambda})$.

We have now a more complex situation here because now we have a family of pdfs $p_1(\mathbf{x}; \vec{\lambda})$.

It turns out that we can still apply the Neyman-Pearson criterion—but now it is applicable to the average detection probability which we denote by $\langle P_D \rangle$, namely, that it is this average detection probability which is maximised for a given false alarm $P_F = \alpha$ (Helstrom [1968]). The average is defined in the following way. Consider a prior or a weight function $w(\vec{\lambda})$ on the parameter space. Without loss of generality, we take it to be normalised:

$$\int_U w(\vec{\lambda}) \, d\vec{\lambda} = 1 \, . \tag{9.6.1}$$

Then the average detection probability is given by,

$$\langle P_D \rangle = \int_U w(\vec{\lambda}) \, P_D(\vec{\lambda}) \, d\vec{\lambda} = \int_U d\vec{\lambda} \, w(\vec{\lambda}) \int_{\mathcal{D}_1} d^N\mathbf{x} \, p_1(\mathbf{x}; \vec{\lambda}) \, . \tag{9.6.2}$$

Then the $\langle P_D \rangle$ is maximised for a given $P_F = \alpha$ for the region $\mathcal{D}_1$ which is now defined via the average likelihood ratio:

$$\langle \Lambda \rangle(\mathbf{x}) = \int_U w(\vec{\lambda}) \, \Lambda(\mathbf{x}; \vec{\lambda}) \, d\vec{\lambda} \, , \tag{9.6.3}$$

where $\Lambda(\mathbf{x}; \vec{\lambda}) = p_1(\mathbf{x}; \vec{\lambda})/p_0(\mathbf{x})$. The regions $\mathcal{D}_1(\Lambda_0)$ are now of the form $\langle \Lambda \rangle(\mathbf{x}) \geq \Lambda_0$. We fix Λ_0 by requiring $P\{\langle \Lambda \rangle(\mathbf{x}) \geq \Lambda_0\} = \alpha$. Then the region $\mathcal{D}_1$ is fixed and the average detection probability $\langle P_D \rangle$ is maximised.

The above discussion fits into the Bayesian framework. The average likelihood ratio $\langle \Lambda \rangle(\mathbf{x})$ is also called the *marginalised* likelihood ratio with respect to a *prior* or a weight function $w(\vec{\lambda})$. We first define conditional probability. Consider two subsets $A, B \subset \Omega$ of the sample space, then the conditional probability is defined as,

$$P(A|B) = \frac{P(A \cap B)}{P(B)} \, . \tag{9.6.4}$$

The conditional probability $P(A|B)$ is the probability of A occurring when B has already occurred. Bayes theorem states:

$$P(A|B) = \frac{p(B|A) \cdot P(A)}{P(B)} \, . \tag{9.6.5}$$

The above equation immediately leads to,

$$P(A|B) \propto p(B|A)\, p(A) \, . \tag{9.6.6}$$

If A is the composite hypothesis H_1 and B is the data $\mathbf{x}$, then $p(A)$ is the prior $w(\vec{\lambda})$ and so the RHS is just the marginalised likelihood ratio $\langle \Lambda \rangle (\mathbf{x})$.

If $w(\vec{\lambda})$ is a slowly varying function of the parameters $\vec{\lambda}$ and the likelihood ratio $\Lambda(\mathbf{x}; \vec{\lambda})$ has a sharp peak at some $\vec{\lambda} = \vec{\lambda}_{max}$, then the dominant contribution to the average likelihood ratio $\langle \Lambda \rangle$ comes from this peak. Therefore we have,

$$\langle \Lambda \rangle (\mathbf{x}) \propto \max_{\vec{\lambda} \in U} \Lambda(\mathbf{x}; \vec{\lambda}) \, . \tag{9.6.7}$$

The detection then can be based on the statistic $\Lambda_{max}(\mathbf{x})$, where,

$$\Lambda_{max}(\mathbf{x}) = \max_{\vec{\lambda} \in U} \Lambda(\mathbf{x}; \vec{\lambda}) \, . \tag{9.6.8}$$

This is called maximum likelihood detection (MLD). Therefore the procedure is as follows. We compute $\Lambda(\mathbf{x}; \vec{\lambda})$ for all $\vec{\lambda} \in U$ and then take the maximum of $\Lambda(\mathbf{x}; \vec{\lambda})$ over $\vec{\lambda}$ and compare this maximum with a pre-assigned threshold to decide the presence or absence of the signal. It must be however be borne in mind that the maximum likelihood is an approximation and its optimality is therefore subject to this consideration.

The maximum likelihood procedure, apart from deciding detection, also estimates the parameters of the signal, that is, the parameters $\vec{\lambda}_{max}$ at which the likelihood ratio $\Lambda(\mathbf{x}; \vec{\lambda})$ is maximised. These are called maximum likelihood estimates (MLE).

9.7 Metric on the Signal Manifold

Sometimes if some of the parameters of the signal are kinematical such as the amplitude, time of coalescence or the phase, they can be easily dealt with in various analytically and computationally efficient ways. If the statistic is a linear function of the data such as the matched filter, we can analytically maximise over the amplitude and also estimate it. We take the compact binary inspiral as a typical example. In order to search over amplitude, we may define a normalised template $\mathbf{s}$ such that $\| \mathbf{s}(\vec{\lambda}) \| = 1$, where now $\vec{\lambda}$ represents all parameters other than the amplitude. These are just normalised matched filters satisfying Eq. (9.4.13) for each value of λ^α. We drop the amplitude from the set of parameters as it will not appear in further discussions involving parameters. Without cause for confusion and to avoid clutter we will continue to denote the rest of the parameters by the same symbol $\vec{\lambda}$. The signal is then $\mathbf{h}(\vec{\lambda}) = A\, \mathbf{s}(\vec{\lambda})$ where A is the amplitude of the signal. The data vector $\mathbf{x}$ is then:

$$\mathbf{x} = \mathbf{h}(\vec{\lambda}) + \mathbf{n} = A\, \mathbf{s}(\vec{\lambda}) + \mathbf{n} \, . \tag{9.7.1}$$

Correlating with $\mathbf{s}(\vec{\lambda})$ we have the SNR ρ,

$$\rho = \langle \mathbf{s}, \mathbf{x} \rangle = A + \langle \mathbf{s}, \mathbf{n} \rangle \,. \tag{9.7.2}$$

The expected value of ρ is $\langle \rho \rangle = A$, since the second term vanishes on taking expectation (recall that the noise has zero mean, that is, $\langle \mathbf{n} \rangle = 0$). Hence, in this specific example, the matched filter output is an unbiased estimator of the amplitude. The point is that we do not need to search over the amplitude. Moreover, we also obtain an unbiased estimate of the amplitude of the signal.

Similarly we can use the fast Fourier transform to search a transient signal over the time of arrival. The phase can be maximised using quadratures (Dhurandhar and Sathyaprakash [1994]; Sathyaprakash and Dhurandhar [1991]) and so on. But there could be other parameters on which the signal depends, which cannot be searched over so easily. For the inspiraling binaries these are the masses, spins etc. Here one needs to densely cover the parameter space with templates—a bank of templates—so that one does not as far as possible miss out any signal. On the other hand since computing resources are finite and limited we have to seek a compromise on the number of templates, so their number must be minimised by placing them as far apart as possible. This number is decided by the maximum mismatch between the signal and the template that can be tolerated. If we are prepared to lose say 10 % of the GW sources, then a maximum mismatch of 3 % in the correlation can be tolerated; a 3 % loss in SNR leads to 10 % loss in the detection volume as the volume scales as the cube of the distance to which we can observe the source. The mismatch is quantified in terms of what is called the *ambiguity function*. We define the ambiguity function as:

$$\mathcal{H}(\vec{\lambda}, \vec{\lambda}') = (\mathbf{s}(\vec{\lambda}), \mathbf{s}(\vec{\lambda}')) \,, \tag{9.7.3}$$

where the signals $\mathbf{s}(\vec{\lambda})$ are normalised as mentioned before and the round brackets denote the scalar product as defined in Eq.(9.3.25). One immediately deduces that $|\mathcal{H}(\vec{\lambda}, \vec{\lambda}')| \leq 1$. This result is immediate from the Schwarz inequality,

$$\mathcal{H}(\vec{\lambda}, \vec{\lambda}') = (\mathbf{s}(\vec{\lambda}), \mathbf{s}(\vec{\lambda}')) \leq \| \mathbf{s}(\vec{\lambda}) \| \; \| \mathbf{s}(\vec{\lambda}') \| = 1 \,. \tag{9.7.4}$$

Since the templates have norm unity, they lie on a submanifold of a hypersphere of $N - 1$ dimensions. The ambiguity function can be thought of as a cosine of the angle between the unit vectors $\mathbf{s}(\vec{\lambda})$ and $\mathbf{s}(\vec{\lambda}')$.

We can now place the templates with the help of the ambiguity function. If the maximum mismatch is taken to be small like 3 % for example, then the template parameters do not differ much, so that we may write, $\vec{\lambda}' = \vec{\lambda} + \Delta\vec{\lambda}$, where $\Delta\vec{\lambda}$ is small. Then we Taylor expand $\mathcal{H}$ to the second order; the first derivative vanishes because when the parameters match, $\mathcal{H}$ attains the maximum value of unity. Thus,

$$\mathcal{H}(\vec{\lambda}, \vec{\lambda} + \Delta\vec{\lambda}) \simeq 1 + \frac{1}{2}\frac{\partial^2 \mathcal{H}}{\partial\lambda^\alpha\partial\lambda^\beta}\Delta\lambda^\alpha\Delta\lambda^\beta,$$

$$\equiv 1 - g_{\alpha\beta}\Delta\lambda^\alpha\Delta\lambda^\beta = 1 - \Delta s^2, \qquad (9.7.5)$$

where we have defined a metric on the parameter space as,

$$g_{\alpha\beta} = -\frac{1}{2}\frac{\partial^2 \mathcal{H}}{\partial\lambda^\alpha\partial\lambda^\beta}. \qquad (9.7.6)$$

This is usually taken as the definition of the metric in the literature. However, geometrically, if $\Delta\theta$ is the angle between $\mathbf{s}(\vec{\lambda})$ and $\mathbf{s}(\vec{\lambda} + \Delta\vec{\lambda})$, then $\mathcal{H} = \cos\Delta\theta \simeq 1 - \Delta\theta^2/2$. And because the angle $\Delta\theta$ is geometrically the distance on a unit (hyper)sphere, we should actually define the metric as:

$$\mathcal{H}(\vec{\lambda}, \vec{\lambda} + \Delta\vec{\lambda}) = 1 - \frac{1}{2}g_{\alpha\beta}\Delta\lambda^a\Delta\lambda^\beta, \qquad (9.7.7)$$

where $g_{\alpha\beta}\Delta\lambda^a\Delta\lambda^\beta$ is now the square of the distance $\Delta\theta$ on the unit hypersphere. Further, this is also the induced metric on the parameter space where the parameter space is regarded as a submanifold of $\mathcal{D}$ on which we already have the metric G_{ik}. See Balasubramanian et al. [1996] where for the first time the metric was defined in the context of GW. Then it is easy to show that,

$$g_{\alpha\beta} = G_{ik}\frac{\partial s^i}{\partial\lambda^\alpha}\frac{\partial s^k}{\partial\lambda^\beta}. \qquad (9.7.8)$$

Note that $s^k(\lambda^\alpha)$ are the components $\mathbf{s}(\lambda^\alpha)$ considered as vectors in $\mathcal{D}$.

However, since this factor of 2 in the definitions does not affect the physical results when used consistently, it may be regarded as a convention. We therefore follow the first formula given in Eq. (9.7.5) so as to be consistent with the literature.

Since some of the parameters, namely, the kinematical parameters like the time of arrival or the phase can be dealt with in a quick manner, the remaining parameters must be searched over with the help of templates. We therefore maximise the match over the kinematical parameters and redefine the ambiguity function over these remaining parameters and call it the reduced ambiguity function (Helstrom [1968]). We will still denote the parameters by $\vec{\lambda}$ with the understanding that now $\vec{\lambda}$ does not include the kinematical parameters (including amplitude) and also continue to denote the reduced ambiguity function by $\mathcal{H}$ and the corresponding metric by $g_{\alpha\beta}$. Maximisation over the kinematical parameters is obtained by minimising the distance Δs defined by Eq. (9.7.5) with respect to the kinematical parameters. See (Owen [1996]) for further details. This then produces another metric on the rest of the parameter space which takes into account the maximisation.

If we call the maximum mismatch as ϵ then we require that $\mathcal{H}(\vec{\lambda}, \vec{\lambda}') \geq 1 - \epsilon \equiv MM$, where then MM is called the minimal match. Thus for the typical case we have considered, $\epsilon = 0.03$ and $MM = 0.97$. The placement of templates is governed by the equation:

$$g_{\alpha\beta} \Delta\lambda^\alpha \Delta\lambda^\beta = \epsilon\,, \tag{9.7.9}$$

where now $g_{\alpha\beta}$ is the metric obtained after maximisation over the kinematical parameters. Typically the contours traced out by $\Delta\lambda^\alpha$ are hyperellipsoids, because $g_{\alpha\beta} \Delta\lambda^\alpha \Delta\lambda^\beta$ is a positive definite quadratic form—this is because $\mathcal{H}$ attains a true maximum when the parameters exactly match. See Owen [1996] for details.

We now present below in the following subsections two applications of the metric.

9.7.1 Placing Templates in the Parameter Space

In order to detect a GW signal one needs to search through the parameter space. The amplitude and kinematical parameters are taken care of easily, but there remain the rest of the parameters, such as, the masses in the case of compact coalescing binaries. Here one needs to scan the parameter space by densely sampling the parameter space. In this endeavour, the metric plays a key role. The problem is really that of "tiling" the parameter space. That is the templates must span the parameter space with the given minimal match, that is leave no "holes" in the parameter space. On the other hand, one must be able to achieve this with a minimum number of templates in order to minimise the computational cost. These are two opposing criteria which govern the placement of templates. Because of the convex shape of the hyper-ellipsoids, there are significant overlaps among the neighbouring templates which must be minimised. So the packing of templates matters. Also there are boundary effects where templates placed near the boundary of the parameter space "spill" outside the boundary—for example the parameter space can become narrow in certain regions, nevertheless templates are required in order to span the space but then they spill out in other directions outside the deemed boundary of the parameter space. Effectively, the dimension of the parameter space can reduce in some part of the parameter space. These considerations affect the number of templates required.

This problem has been comprehensively addressed in (Allen [2021]) and (Allen and Shoom [2021]). A simple strategy is to arrange them in the form of a hypercube. But this is not optimal. In two dimensions a hexagonal packing reduces the overlap significantly. This reduces the number of templates by about 23 % as compared with the placement on the vertices of a square. In the Fig. 9.3 we show how templates can be packed in a hexagonal pattern in the case of spinless compact inspiraling binary signals. The parameters are the Newtonian and post-Newtonian chirp times τ_0 and τ_3 which are certain functions of the two masses m_1 and m_2 (Sengupta et al. [2002]).

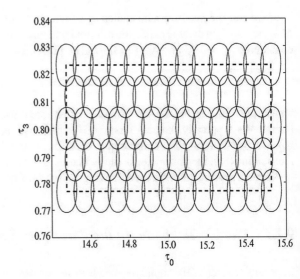

Fig. 9.3 An hexagonal packing of templates for the nonspinning binary inspiral parameter space. Hexagonal packing is more efficient than square packing by about 23 %. Figure source: Sengupta et al. [2002], Class. Quantum Grav. 19, 1507 (2002). Copyright IOP Publishing. Reproduced with permission. All rights reserved

9.7.2   Fisher Information Matrix and Rao-Cramer Bound

As we saw in the previous section, maximum likelihood method also furnishes estimates of the parameters of the signal. These are called maximum likelihood estimates or MLE in short. We will denote these estimates of the parameters of the signal by $\vec{\lambda}_{\text{MLE}}$. However, the true values of the parameters in general will differ from the MLE estimates because of the presence of noise. Therefore, errors are introduced in the estimation process. The question is whether we can in anyway quantify these errors. The answer to this question is yes—but we can do so only statistically. The Rao-Cramer bound does just this; it puts bounds on the covariance of the errors and this is done in terms of the metric defined earlier. Let $\Delta\vec{\lambda} = \vec{\lambda}_{\text{true}} - \vec{\lambda}_{\text{MLE}}$, then the Rao-Cramer bound says:

$$\langle\, \Delta\lambda^\alpha\, \Delta\lambda^\beta\,\rangle \geq \frac{1}{A^2}\, g^{\alpha\beta}\,. \tag{9.7.10}$$

Here $g^{\alpha\beta}$ is the inverse of the metric $g_{\alpha\beta}$, that is, $g^{\alpha\gamma}\, g_{\gamma\beta} = \delta^\alpha_\beta$. We see from Eq. (9.7.10) that the errors reduce as the amplitude A of the signal increases or as the signal becomes stronger as is expected. We now sketch the proof of the Rao-Cramer inequality which in our ideal situation reduces to an equality.

We write $l(\mathbf{x}; \vec{\lambda}) = \ln \Lambda(\mathbf{x}; \vec{\lambda})$ and consider its derivate with respect λ^α and write it as $l_{,\alpha}$. This is called the score. We note the following properties of $l_{,\alpha}$.

$$\langle l_{,\alpha}\,\rangle = \left\langle \frac{p_{1,\alpha}}{p_1} \right\rangle = \frac{\partial}{\partial\lambda^\alpha}\left(\int p_1 d\mathbf{x}\right) = 0$$

$$\langle l_{,\alpha\beta}\,\rangle = -\langle l_{,\alpha}\, l_{,\beta}\,\rangle\,. \tag{9.7.11}$$

The second equation follows from:

$$l_{,\alpha\beta} = \frac{p_{1,\alpha\beta}}{p_1} - \left(\frac{p_{1,\alpha}}{p_1}\right)\left(\frac{p_{1,\beta}}{p_1}\right). \tag{9.7.12}$$

On taking the expected value of the above identity, the first term vanishes and the second term reproduces the RHS in Eq. (9.7.11). We see that the expected value of the score is zero at the maximum from the first equation and its covariance $g_{\alpha\beta}^{\text{Rao}-\text{Fisher}} = \langle l_{,\alpha} \, l_{,\beta} \rangle$ is called the Fisher information matrix. Geometrically, the score $l_{,\alpha}$ can be thought of a gradient of the hypersurface defined by $l = \text{const.}$ and then the Fisher matrix is essentially the curvature of the surface. Thus, larger the curvature, the sharper is the peak, that is, bigger the Fisher information matrix, the sharper the maximum and more accurate the MLE. The Fisher matrix can also be viewed as a metric on the family of probability distributions. This is the Rao-Fisher metric. It is a measure of how much the probability distributions differ when we change from λ^{α} to $\lambda^{\alpha} + \Delta\lambda^{\alpha}$.

The next step is to Taylor expand $l_{,\alpha}$ about $\vec{\lambda}_{\text{true}}$ to first order and set it equal to zero at the $\vec{\lambda}_{\text{MLE}}$. We then have:

$$l_{,\alpha} + \Delta\lambda^{\beta} \, l_{,\alpha\beta} \simeq 0, \tag{9.7.13}$$

where $\Delta\lambda^{\alpha} = \lambda^{\alpha}_{\text{MLE}} - \lambda^{\alpha}_{\text{true}}$. This equation then gives us:

$$\Delta\lambda^{\beta} \, l_{,\alpha\beta} = -l_{,\alpha} . \tag{9.7.14}$$

We next replace $l_{,\alpha\beta}$ by its expected value and use Eq. (9.7.11). We now assume that the noise is Gaussian. Then $l = A(\mathbf{x}, \mathbf{s}) - 1/2A^2$. For the time being and to avoid clutter we set $A = 1$; we will insert it in the last step. Then we have the following relations:

$$l_{,\alpha} = (\mathbf{n}, \mathbf{s}_{,\alpha}),$$
$$\langle l_{,\alpha} \, l_{,\beta} \rangle = \langle (\mathbf{n}, \mathbf{s}_{,\alpha})(\mathbf{n}, \mathbf{s}_{,\beta}) \rangle,$$
$$\equiv (\mathbf{s}_{,\alpha}, \mathbf{s}_{,\beta}) = g_{\alpha\beta}. \tag{9.7.15}$$

Thus the Fisher information matrix is equivalent to the metric we derived using the ambiguity function. The middle step in the above equation is derived in Exercise 3. Therefore using Eqs. (9.7.11, 9.7.14 9.7.15) we arrive at:

$$\Delta\lambda^{a} = g^{\alpha\beta} \, l_{,\beta} . \tag{9.7.16}$$

This immediately gives the result that,

$$\langle \Delta\lambda^{a} \, \Delta\lambda^{\beta} \rangle = g^{\alpha\gamma} g^{\beta\delta} \langle l_{,\gamma} l_{,\delta} \rangle = g^{\alpha\gamma} g^{\beta\delta} g_{\gamma\delta} = g^{\alpha\beta} \tag{9.7.17}$$

Here we have obtained an equality. However, in the general case we have an inequality—the Rao-Cramer inequality. Inserting the amplitude A back into the equations we obtain:

$$\langle \Delta\lambda^a \; \Delta\lambda^\beta \rangle \; \geq \; \frac{1}{A^2} \, g^{\alpha\beta} \,, \tag{9.7.18}$$

which gives a bound on the covariance of the errors in the MLE. The inequality sign between square matrices M_1 and M_2, that is, $M_1 \geq M_2$ means that the matrix $M_1 - M_2$ is positive semi-definite or equivalently it has non-negative eigenvalues. The diagonal terms $g^{\alpha\alpha}/A^2$ give the bounds on the variances of the parameters λ^α. The inequality tends to an equality in the limit of large SNR.

In Exercise 4 we outline the proof the Rao-Cramer inequality in general for a single parameter family of probability distributions. For the general case of several parameters, we refer the reader for a discussion and proof to Papoulis and Pillai [2002].

9.8   Burst and Stochastic Searches

Searches which employ matched filtering, by definition, require a modelled waveform for a signal. This is however possible only in specific cases of coalescing binaries and asymmetric spinning compact stars. One however expects various known and unknown astrophysical events to produce gravitational wave signals, whose waveform cannot be obtained with certainty, though the energy emission profile can be modeled. Such sources, when short in duration, are called burst sources. Supernovae, cosmic strings are examples of such sources. To search for these sources, which would produce a tiny excess power in the detector output, typically one looks for similar patterns or coherence in multiple detectors. Even though it may not be possible to model a signal, the phase of the signal can be correlated across different detectors—one detector's data can be correlated with another detector's data. One can use this fact to eliminate the chances of coincident noise transients in the detectors. Even a binary merger can be considered as a burst source and a model independent search can be carried for these sources. This can help us to find out if there are signals whose waveforms differ from those based on accepted models of the sources.

A stochastic background is a persistent unmodelled source. It is produced by several independent unresolved sources. It can be searched using a cross-correlation algorithm. Here, however, one can further use earth rotation synthesis imaging, similar to radio astronomy, to make an image of the sky. This algorithm is known as gravitational wave radiometer. While no stochastic background has been detected so far, upperlimits on isotropic and anisotropic backgrounds are routinely obtained from the data of groundbased detectors. See Romano and Cornish [2017] for a review.

9.9 Present Challenges

Gravitational wave astronomy has just begun. It was an exciting beginning with the detection fairly large number of GW events. Different kinds of binary mergers—black hole—black hole, neutron star—neutron star, black hole—neutron star—were observed in the first 5 years. However, a long road lies ahead in the future, as can be imagined from the progress of electromagnetic astronomy over past four centuries. Here we have made enormous progress since Galileo turned the telescope to the sky for observing astronomical objects 400 years ago. Though sensitivity of the detectors needs to be increased and next generation detectors must be built to achieve this goal, data analysis algorithms can make significant contributions to the overall objectives.

For every detected binary merger, there are millions of instrumental and environmental transients, a.k.a. glitches, in the data. Sophisticated mathematical and modern computational techniques like machine learning are being developed to tackle these issues.

In the near future, the primary goal will be to detect a large number of compact binary coalescences, which will help us in understanding their origin, evolution and distribution in the universe. However, the most interesting binaries—those precessing—the ones where the component masses do not have spins aligned with the orbital angular momentum are difficult to detect with the present algorithms. The dimension of the parameter space must be increased, requiring millions of templates. This search is not feasible with the presently available computing power. Smart algorithms will be necessary to achieve this.

GW astronomy is expected to observe different kind of sources, not just binary mergers. Each of these sources pose their own detection challenges and innovative data analysis techniques will be needed to look for these. Searches for continuous wave sources are limited by the computational cost demanded by the high frequency resolution. A group theoretic approach which exploits symmetries in the problem may provide a solution. Searches for burst sources, like supernovae, may benefit from machine learning. Searches for stochastic background may benefit from introduction of advanced techniques to characterise a field defined on a two-sphere. Finally, detection using space-based detectors requires techniques like time-delay interferometry (Tinto and Dhurandhar [2005]).

Gravitational wave astronomy therefore has a lot to offer. It promises interesting sources that may be observed in the near future. Their detection remains intellectually challenging, which perhaps makes the field even more interesting!

So far we have barely observed the tip of the iceberg.

Exercises

1. Consider the multivariate Gaussian pdf for the random vector X defined by:

$$p(\mathbf{x}) = \frac{1}{(2\pi)^{N/2}\sqrt{\det C}}\, e^{-\frac{1}{2}\mathbf{x}^T C^{-1}\mathbf{x}}. \tag{9.9.1}$$

(a) Show by direct integration that $\langle x_i x_k \rangle = \int x_i x_k \, p(\mathbf{x}) \, d\mathbf{x} = C_{ik}$.
(b) An efficient method is to use the moment generating function $M(\xi) = \int d\mathbf{x} \, p(\mathbf{x}) \, e^{\xi^T \cdot \mathbf{x}}$ where $\xi = (\xi_0, \xi_1, ..., \xi_{N-1})^T$ is a column vector. Show that $M(\xi) = e^{\frac{1}{2} \xi^T \cdot C \cdot \xi}$ and hence obtain:

$$\langle x_i x_k \rangle = \left[\frac{\partial^2 M}{\partial \xi_i \, \partial \xi_k} \right]_{\xi=0} \equiv C_{ik}.$$

2. Maximise $\langle \rho \rangle(q)$ in Eq. (9.4.15) with respect q subject to the normalising condition Eq. (9.4.13) by using Lagrange mulipliers and thus obtain the matched filter. Write:

$$F(q) = \int \tilde{q}^*(f) \, \tilde{h}(f) \, df - \lambda \left(\int df \, |\tilde{q}(f)|^2 S(f) - 1 \right), \qquad (9.9.2)$$

where λ is a Lagrange multiplier. Note that here we have a *functional* $\langle \rho \rangle(q)$ which must be maximised with respect to a function q.
(Hint: An easy way is to discretise the above equation and write it as:

$$F(q_i) = q_i^* h_i - \lambda (q_i q_i^* S_i - 1),$$

where summation is implied over repeated index. Now differentiate w.r.t. q_k^*,

$$\frac{\partial F}{\partial q_k^*} = h_k - \lambda q_k \, S_k \equiv 0, \qquad (9.9.3)$$

which gives the result $q_k \propto h_k / S_k$. This equation can now be appropriately interpreted in the continuous domain.)
3. Given two data vectors $\mathbf{x}, \mathbf{y} \in \mathcal{D}$, and noise $\mathbf{n}$ satisfying Eq. (9.3.14) show that,

$$\langle (\mathbf{x}, \mathbf{n}) \, (\mathbf{n}, \mathbf{y}) \rangle = (\mathbf{x}, \mathbf{y}).$$

(Hint: Use the steps similar to those used in deriving Eq. (9.4.10).)
4. Prove the Rao-Cramer inequality when the family of signals depends on a single parameter θ. Let $\hat{\theta}$ be an unbiased estimator of θ, that is,

$$\int (\hat{\theta} - \theta) \, p(x, \theta) \, dx = 0,$$

for all θ. Then its derivative with respect to θ is also zero. Hence show that:

$$\int (\hat{\theta} - \theta) \, p \, l_{,\theta} \, dx = 1,$$

where $l = \ln p(x, \theta)$. Now use the Schwarz inequality to show that:

$$\int (\hat{\theta} - \theta)^2 \, p \, dx \, \int (l_{,\theta})^2 \, p \, dx \geq 1,$$

and hence show that:

$$\langle (\hat{\theta} - \theta)^2 \rangle \geq \langle (l_{,\theta})^2 \rangle^{-1}.$$

Correction to: General Relativity and Gravitational Waves

Correction to:
S. Dhurandhar and S. Mitra, *General Relativity*
and Gravitational Waves, **UNITEXT for Physics,**
https://doi.org/10.1007/978-3-030-92335-8

The original version of the book has received belated corrections in some equations in Chapters 1, 4, 8 and 9, which have now been incorporated. The chapters have been updated with the changes.

The updated original version of these chapters can be found at
https://doi.org/10.1007/978-3-030-92335-8_1
https://doi.org/10.1007/978-3-030-92335-8_4
https://doi.org/10.1007/978-3-030-92335-8_8
https://doi.org/10.1007/978-3-030-92335-8_9

C1

Bibliography

Adhikari, R. X. (2014). Gravitational radiation detection with laser interferometry. *Reviews of Modern Physics, 86*, 121–151.

Aguiar, O. D. (2011). Past, present and future of the Resonant-Mass gravitational wave detectors. *Research in Astronomy and Astrophysics, 11*, 1–42.

Allen, B. (2021). Optimal Template Banks.

Allen, B. and Shoom, A. A. (2021). Template banks based on Z^n and A_n^* lattices.

Ashtekar, A. and Bonga, B. (2017). On the ambiguity in the notion of transverse traceless modes of gravitational waves. General Relativity and Gravitation *49*.

Balasubramanian, R., Sathyaprakash, B. S., & Dhurandhar, S. V. (1996). Gravitational waves from coalescing binaries: Detection strategies and Monte Carlo estimation of parameters. *Phys. Rev. D, 53*, 3033–3055.

Baumgarte, T. W., & Shapiro, S. L. (2010). *Numerical Relativity: Solving Einstein's Equations on the Computer.* Cambridge University Press.

Blanchet, L. (2002). Gravitational Radiation from Post-Newtonian Sources and Inspiralling Compact Binaries. Living Reviews in Relativity *5*.

Blanchet, L., Iyer, B. R., & Joguet, B. (2002). Gravitational waves from inspiraling compact binaries: Energy flux to third post-Newtonian order. *Phys. Rev. D, 65*, 064005.

Chandrasekhar, S. (2002). The mathematical theory of black holes. Oxford classic texts in the physical sciences, Oxford Univ. Press, Oxford.

Dhurandhar, S. V., & Sathyaprakash, B. S. (1994). Choice of filters for the detection of gravitational waves from coalescing binaries. 2. Detection in colored noise. *Phys. Rev. D, 49*, 1707–1722.

Dhurandhar, S. V., & Tinto, M. (1988). Astronomical observations with a network of detectors of gravitational waves. I - Mathematical framework and solution of the five detector problem. *Month. Not. Roy. Astron. Soc., 234*, 663–676.

Eisenhart, L. P. (1926). *Riemannian Geometry.* Princeton, USA: Princeton University Press.

Flanagan, É. É., & Hughes, S. A. (2005). The basics of gravitational wave theory. *New Journal of Physics, 7*, 204.

Gelfand, I. M., Minlos, R. A., & Shapiro, S. . Z. . Ya. . (1963). *Representations of the Rotation and Lorentz Groups and Their Applications.* Oxford: Pergamon Press.

Goldberg, J. N., Macfarlane, A. J., Newman, E. T., Rohrlich, F., & Sudarshan, E. C. G. (1967). Spin-s Spherical Harmonics and & ð. *Journal of Mathematical Physics, 8*, 2155–2161.

© The Editor(s) (if applicable) and The Author(s), under exclusive license to Springer Nature Switzerland AG 2022
S. Dhurandhar and S. Mitra, *General Relativity and Gravitational Waves*, UNITEXT for Physics, https://doi.org/10.1007/978-3-030-92335-8

Goldstein, H. (1980). *Classical Mechanics* (2nd ed.). Reading, MA: Addison-Wesley Publishing Company.

Griffiths, D. (2012). *Introduction to Electrodynamics* (4th ed.). Reading, MA: Addison Wesley.

Hartle, J. (2003). *Gravity: An Introduction to Einstein's General Relativity*. Reading, MA: Addison-Wesley.

Helstrom, C. W. (1968). *Statistical Theory of Signal Detection*. Pergamon.

Hough, J., & Rowan, S. (2000). *Living Reviews in Relativity, 3*, 3.

Isaacson, R. A. (1968). Gravitational Radiation in the Limit of High Frequency. II. Nonlinear Terms and the Effective Stress Tensor. *Phys. Rev., 166*, 1272–1280.

Jackson, J. D. (1998). *Classical Electrodynamics*. New York: Wiley.

Kembhavi, A., & Khare, P. (2020). *Gravitational Waves: A New Window to the Universe*. Singapore: Springer.

Kerr, R. P. (1963). Gravitational Field of a Spinning Mass as an Example of Algebraically Special Metrics. *Phys. Rev. Lett., 11*, 237–238.

Kruskal, M. D. (1960). Maximal Extension of Schwarzschild Metric. *Phys. Rev., 119*, 1743–1745.

Landau, L. D., & Lifshitz, E. M. (1980). *Classical Theory of Fields, vol. 2 of Course of Theoretical Physics*. Oxford, UK: Pergamon Press.

Martynov, D. V., et al. (2016). Sensitivity of the Advanced LIGO detectors at the beginning of gravitational wave astronomy. *Phys. Rev. D, 93*, 112004.

Mathews, J., & Walker, R. L. (1970). *Mathematical methods of physics*. New York: Addison-Wesley.

Misner, C. W., Thorne, K. S., & Wheeler, J. A. (1973). *Gravitation*. New York: Freeman.

Narlikar, J. V. (2010). *An Introduction to Relativity*. Cambridge, UK: Cambridge University Press.

Owen, B. J. (1996). Search templates for gravitational waves from inspiraling binaries: Choice of template spacing. *Phys. Rev. D, 53*, 6749–6761.

Padmanabhan, T. (1993). *Structure Formation in the Universe*. New York: Cambridge University Press.

Padmanabhan, T. (2010). *Gravitation: Foundations and Frontiers* (1st ed.). New York: Cambridge University Press.

Papoulis, A., & Pillai, S. U. (2002). *Probability, Random Variables, and Stochastic Processes* (4th ed.). Boston: McGraw Hill.

Peters, P. C., & Mathews, J. (1963). Gravitational Radiation from Point Masses in a Keplerian Orbit. *Phys. Rev., 131*, 435–440.

Press, W. H., Teukolsky, S. A., Vetterling, W. T., & Flannery, B. P. (2007). *Numerical Recipes: The Art of Scientific Computing* (3rd ed.). New York: Cambridge University Press.

Racz, I. (2009). Gravitational radiation and isotropic change of the spatial geometry.

Raychaudhuri, A. K., & Sriranjan Banerji, A. B. (1992). *General relativity, astrophysics, and cosmology*. Berlin: Springer.

Romano, J. D., & Cornish, N. J. (2017). Detection methods for stochastic gravitational-wave backgrounds: a unified treatment. *Living Reviews in Relativity, 20*, 2.

Sathyaprakash, B. S., & Dhurandhar, S. V. (1991). Choice of filters for the detection of gravitational waves from coalescing binaries. *Phys. Rev. D, 44*, 3819–3834.

Sathyaprakash, B. S., & Schutz, B. F. (2009). Physics, Astrophysics and Cosmology with Gravitational Waves. *Living Reviews in Relativity, 12*, 2.

Saulson, P. R. (2017). *Fundamentals of Interferometric Gravitational Wave Detectors* (2nd ed.). Singapore: World Scientific.

Schutz, B. F. (1995). *A first course in general relativity*. Cambridge, UK: Cambridge University Press.

Scientific Collaboration, L. I. G. O., Collaboration, Virgo, Abbott, B. P., et al. (2016). Observation of Gravitational Waves from a Binary Black Hole Merger. *Phys. Rev. Lett., 116*, 061102.

Scientific Collaboration, L. I. G. O., Collaboration, Virgo, Abbott, R., et al. (2021). GWTC-2: Compact Binary Coalescences Observed by LIGO and Virgo during the First Half of the Third Observing Run. *Physical Review X, 11*, 021053.

Scientific Collaboration, L. I. G. O., Collaboration, Virgo, Collaboration, K. A. G. R. A., et al. (2021b). GWTC-3: Compact Binary Coalescences Observed by LIGO and Virgo During the Second Part of the Third Observing Run. *General Relativity and Quantum Cosmology, 2111*, 03606.

Sengupta, A. S., Dhurandhar, S. V., Lazzarini, A., & Prince, T. (2002). Extended hierarchical search (EHS) algorithm for detection of gravitational waves from inspiralling compact binaries. *Classical and Quantum Gravity, 19*, 1507–1512.

Spiegel, M. R. (1959). *Vector Analysis and an Introduction to Tensor Analysis. Schaum Outline Series*. London: McGraw-Hill.

Tinto, M., & Dhurandhar, S. V. (2005). Time-delay interferometry. *Living Reviews in Relativity, 8*(1), 4.

Touboul, P., et al. (2017). MICROSCOPE Mission: First Results of a Space Test of the Equivalence Principle. *Phys. Rev. Lett., 119*, 231101.

Vishveshwara, C. V. (1968). Generalization of the & "Schwarzschild Surface" to arbitrary static and stationary metrics. *J. Math. Phys., 9*, 1319.

Vishveshwara, C. V. (1970). Scattering of gravitational radiation by a Schwarzschild black hole. *Nature, 227*, 936–938.

Weinberg, S. (1972). *Gravitation and Cosmology: Principles and applications of the general theory of relativity*. New York: Wiley.

Index

Printed in the United States
by Baker & Taylor Publisher Services